AF441845

Gas Dynamics

Gas Dynamics

Theory and Applications

George Turrell
Université des Sciences et Technologies de Lille, Villeneuve d'Ascq, France

JOHN WILEY & SONS

Chichester · New York · Weinheim · Brisbane · Singapore · Toronto

Other Wiley Editorial Offices

John Wiley & Sons, Inc., 605 Third Avenue,
New York, NY 10158-0012, USA

VCH Verlagsgesellschaft mbH, Pappelallee 3,
D-69469 Weinheim, Germany

Jacaranda Wiley Ltd, 33 Park Road, Milton,
Queensland 4064, Australia

John Wiley & Sons (Asia) Pte Ltd, 2 Clementi Loop #02-01,
Jin Xing Distripark, Singapore 129809

John Wiley & Sons (Canada) Ltd, 22 Worcester Road,
Rexdale, Ontario M9W 1L1, Canada

Library of Congress Cataloging-in-Publication Data

Turrell, George.
    Gas dynamics : theory and applications / George Turrell.
       p.   cm.
    Includes bibliographical references and index.
    ISBN 0-471-97573-7 (cloth : alk. paper).
    1. Gas dynamics. I. Title.
    QC168.T92   1997   97-7328
    530.4′3—dc21   CIP

British Library Cataloguing in Publication Data

A catalogue record for this book is available from the British Library

ISBN 0 471 97573 7 (cloth)

Typeset in 10/12pt Times by Keytec Typesetting Ltd, Bridport, Dorset
Printed and bound in Great Britain by Biddles Ltd, Guildford, Surrey
This book is printed on acid-free paper responsibly manufactured from sustainable forestation, for which at least two trees are planted for each one used for paper production

Contents

Preface

This little book, which I have nicknamed 'Gaston', was born many years ago when I was asked to teach a short course at the University of Bordeaux on the properties of gases. The objective was to interest students at the level of advanced undergraduates in the study of the gaseous state. The concept of the 'ideal gas' was of course already well known to them. Thus, the logical development of the theory was to introduce a description of intermolecular forces and their role in determining the properties of real gases, including the transport properties.

For experimentalists in this field, the period was an exciting one. The results of some of the earlier flash-photolysis experiments had only recently been reported, shock-tube studies of chemical reactions were in their prime and the idea of crossed molecular-beam experiments was being formulated. The entire subject of real-time investigation of physical and chemical phenomena was just getting underway. Then, the first lasers appeared on the scene and the possibilities of research in gas-phase dynamics became enormous.

Over the years Gaston matured with the subject and served as a basis for courses which I taught in the Zaïre and Québec, as well as in France. The audience varied, including students in physical chemistry, physics and chemical engineering. Although the examples chosen and their presentation were adjusted accordingly, the courses consisted essentially of two parts, theory and applications. I have retained this structure here.

Gaston has been put to bed with the help of many, including 'Miss Mac' for the preparation of the text and figures. I should like to thank my colleague Daniel Couturier and his research group for their aid when she was in her more stubborn moods. My very special thanks are to Irène, for her patience when Gaston was particularly trying.

Lille, April 1997   George Turrell

Part I
BASIC THEORY

Chapter 1

The Elementary Kinetic Theory of Gases

This kinetic theory is developed by considering a single molecule in a box, a rectangular parallelepiped of dimensions a, b, c, as shown in Fig. 1. The velocity of the molecule is described by a vector $\mathbf{u}$, such that the speed u is given by

$$u^2 = \dot{x}^2 + \dot{y}^2 + \dot{z}^2, \tag{1}$$

where in the notation of Newton, the dot over a variable indicates that it is preceded by the operator $\mathrm{d}/\mathrm{d}t$, where t is the time. Thus, $\dot{x}$, $\dot{y}$ and $\dot{z}$ are the components of the velocity of the molecule along the axes shown in Fig. 1. Each collision that the molecule makes with a wall of the box is assumed to be elastic in the sense that its kinetic energy does not change. Therefore, although the direction of its motion is reversed as a result of the collision with the wall, the magnitude of its momentum is conserved.

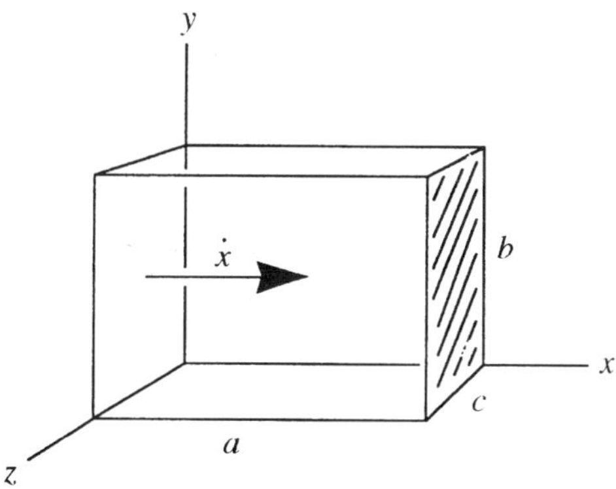

Fig. 1   Kinetic theory model of the collision of a molecule with a wall

If the molecule makes Z collisions per second with a given wall in the y, z plane, then the absolute value of the change in the x component of the velocity is given by

$$|\dot{x}| = 2aZ, \tag{2}$$

as $2a$ represents the total distance that the molecule must travel between successive collisions with the same wall. Each time the molecule strikes the wall its momentum p changes by

$$\Delta p = 2m|\dot{x}|, \tag{3}$$

where m is the mass of the molecule. Then, from Eq. (2) the rate of change of momentum becomes equal to

$$\frac{\Delta p}{\Delta t} = 2m|\dot{x}|Z = \frac{m\dot{x}^2}{a}. \tag{4}$$

Now assume that the box is filled with a large number, say N, of identical molecules, and that these molecules do not interact with each other. In other words, it is assumed here that intermolecular forces are negligible and, furthermore, that the diameter of a molecule is negligible compared with the average distance between two molecules. The latter assumption is equivalent mathematically to a model in which each molecule is represented by a point mass. For the moment, the notion of an internal structure of a molecule is not introduced.

Finally, the molecules in the box are considered to be in thermal motion, with a kinetic energy that is proportional to an absolute temperature. For the purpose of the present derivation, the temperature will be held constant.

If all the molecules in the box are identical, then the total rate of change of momentum is given by

$$\sum_i \frac{\Delta p_i}{\Delta t} = \frac{1}{a}\sum_i m_i\dot{x}_i^2 = \frac{m}{a}\sum_i \dot{x}_i^2. \tag{5}$$

From Newton's second law of motion, Eq. (5) represents the total force acting on the wall as a result of molecular collisions (see Fig. 1).

The mean-square velocity in the x direction is equal to

$$\overline{\dot{x}^2} = \sum_i \frac{\dot{x}_i^2}{N} \tag{6}$$

and the force on the wall of area A can be expressed as

$$\sum_i \frac{\Delta p_i}{\Delta t} = \frac{mN}{a}\overline{\dot{x}^2} = Pbc = PA. \tag{7}$$

The pressure P is the force per unit area on the wall. Thus, Eq. (7) yields the relation

$$PV = mN\overline{\dot{x}^2}. \tag{8}$$

If, following Boltzmann, the three directions x, y and z are assumed to be equivalent [see Eq. (1)], then

$$\overline{u^2} = 3\overline{\dot{x}^2} \tag{9}$$

and

$$PV = \tfrac{1}{3}Nm\overline{u^2}. \tag{10}$$

The total kinetic energy of the system is then given by

$$E = \tfrac{1}{2}\sum_i m_i\overline{u^2} = \tfrac{1}{2}Nm\overline{u^2} \tag{11}$$

for the N identical particles of mass m.

The molecular kinetic energy of the system will now be used as the basis of a temperature scale. Thus, the temperature will be defined by

$$E = NCT, \tag{12}$$

where C, a constant of proportionality, will be determined below. From Eqs (10)–(12)

$$E = \tfrac{3}{2}PV = NCT \tag{13}$$

or

$$PV = \tfrac{2}{3}NCT. \tag{14}$$

It is convenient to define the 'gas constant', R, by the relation

$$R = \tfrac{2}{3}CN_{\mathrm{o}}, \tag{15}$$

where $N_{\mathrm{o}} = 6.022 \times 10^{23}$ (molecules per mole) is Avogadro's number. The combination of Eqs (14) and (15) yields

$$PV = n_{\mathrm{o}}RT, \tag{16}$$

where $n_{\mathrm{o}} = N/N_{\mathrm{o}}$ is the number of moles of gas in the system. The constant R is of great importance. It can be expressed in several different systems of units, as shown in Table 1.

Table 1 Gas constant, R

Units	Value
$\mathrm{J\,K^{-1}\,mol^{-1}}$	8.314
$\mathrm{cal\,K^{-1}\,mol^{-1}}$	1.987
$\mathrm{liter\,atm\,K^{-1}\,mol^{-1}}$	0.082 06
$\mathrm{cm^3\,atm\,K^{-1}\,mol^{-1}}$	82.06

With the use of Eqs (10) and (16), the root-mean-square molecular velocity can be written as

$$\sqrt{\overline{u^2}} = \sqrt{\frac{3PV}{N_o m}} = \sqrt{\frac{3PV}{M}},\tag{17}$$

where $M = mN_o$ is the molecular weight of the gas.

It is important to summarize the approximations made in the simple model developed above, namely

   (i)   there are no forces acting between the molecules,
  (ii)   the diameter of each molecule is negligible compared with the distances between the molecules,
 (iii)   the molecules are in thermal agitation with a total energy that is proportional to an absolute temperature, and
 (iv)   all collisions between the molecules, as well as all collisions with the walls of the box, are elastic, e.g. no energy is exchanged.

The above approximations, which lead to Eq. (16), serve as the definition of an ideal gas. In this case, the quantity $PV/n_o = RT$ should be independent of both the nature of the gas and the pressure P. However, for real gases, this quantity varies with the pressure, as shown by the examples in Fig. 2. At a given temperature, PV/n_o has the same value for all gases, in the limit as $P \rightarrow 0$. At this point, all of the approximations listed above become valid and the gases can be considered to be ideal. This property provides the possibility of defining a scale of absolute temperature, i.e. the identification by means of Eq. (15) of the constant of proportionality C introduced in Eq. (12).

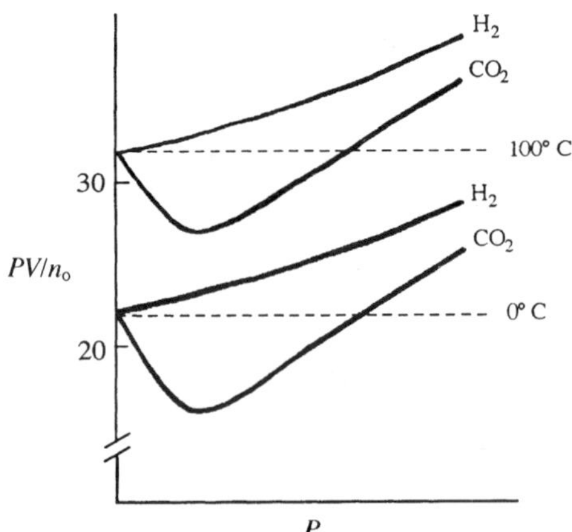

Fig. 2   Schematic representation of PV/n_o vs. P for hydrogen and carbon dioxide; the dotted lines represent the behavior of an ideal gas

The temperature dependence of the values of PV/n_0 as $P \to 0$ can be expressed in the form

$$\lim_{P \to 0} \left(\frac{PV}{n_0} \right) = \mathrm{a} + \mathrm{b}t \tag{18}$$

$$= \left(\frac{PV}{n_0} \right)_0 = RT, \tag{19}$$

where, in Eq. (18), t is the temperature on the centigrade scale. Note that the centigrade scale is defined by the freezing and boiling points of water at atmospheric pressure, 0 °C and 100 °C, respectively (see Fig. 3). The origin of the absolute temperature scale can be obtained by extrapolation to the point at which the kinetic energy vanishes. The Kelvin scale of absolute temperature is thus determined and the kinetic energy can be expressed in the form

$$E = CNT = \tfrac{3}{2}PV = \tfrac{3}{2}n_0 RT. \tag{20}$$

From thermodynamics, the definition of the heat capacity of a gas at constant volume is given by

$$C_V = \left(\frac{\partial E}{\partial T} \right)_V. \tag{21}$$

For the system of particles, structureless molecules, considered here in three dimensions, Eq. (20) yields

$$C_V = \tfrac{3}{2}n_0 R \tag{22}$$

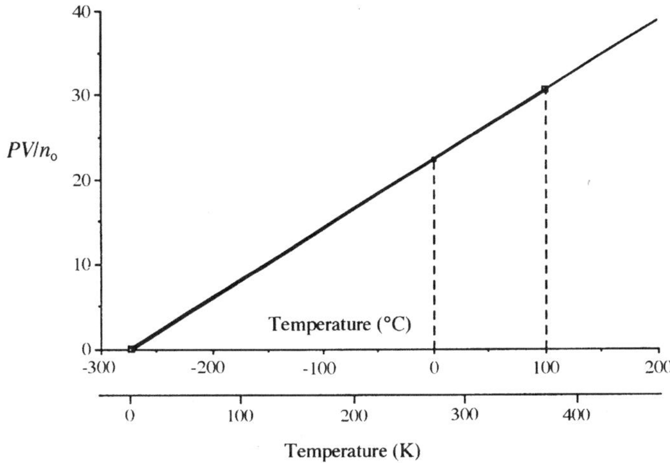

Fig. 3   Determination of the absolute temperature scale by extrapolation

or

$$\tilde{C}_V = \frac{C_V}{n_0} = \tfrac{3}{2}R \tag{23}$$

per mole of gas. As the three directions in space are equivalent, it is evident that Eq. (23) corresponds to a contribution to $\tilde{C}_V$ of $\tfrac{1}{2}R$ for each direction, or translational degree of freedom of the system. The extension of this analysis to include molecular structure, i.e. the existence of the internal degrees of freedom of molecular rotation and vibration, will be made in Chapter 4. There, it will be demonstrated that these additional degrees of freedom also contribute to the value of the heat capacity of the gas.

Chapter 2

Molecular Statistics

According to quantum theory, the stationary state of a system is defined by a wave function, traditionally represented by the symbol ψ. Each state corresponds to a discrete value of the energy, an *energy level*. If there is only one state—one wave function—for each energy level, the level is said to be non-degenerate. If, however, there is more than one wave function that corresponds to the same energy, then the system is degenerate. In this case, the degree of degeneracy is equal to the number of independent wave functions associated with the same energy level.

In the development of molecular statistics it is necessary to specify the distribution of molecules over the various energy levels of a system. For this purpose, it is useful to introduce the notion of probability, which for a given event can be defined by the relation

$$\mathscr{W} = \frac{n}{\mathscr{N}},\tag{24}$$

where n is the number of favorable results and $\mathscr{N}$ is the number of possible results.

As a simple example, consider a deck of 52 cards. According to Eq. (24), the probability of drawing a heart from the deck is given by

$$\mathscr{W}(\heartsuit) = \tfrac{13}{52} = \tfrac{1}{4},\tag{25}$$

as there are, of course, 13 hearts in the deck. Obviously, the same result is obtained for the probability of drawing, say, a spade. The probability of drawing either a heart or a spade is then given by the sum

$$\mathscr{W}(\heartsuit) + \mathscr{W}(\spadesuit) = \tfrac{1}{4} + \tfrac{1}{4} = \tfrac{1}{2},\tag{26}$$

because they are independent results. However, note that the combined probability of drawing the ace of spades is equal to

$$\mathscr{W}(\mathscr{A})\mathscr{W}(\spadesuit) = \tfrac{1}{13} \times \tfrac{1}{4} = \tfrac{1}{52}.\tag{27}$$

Consider now the a system composed of N distinguishable particles. Assume that there are N_1 particles in energy level ε_1, N_2 in energy level ε_2, etc. The probability of a given distribution over the ensemble of energy levels is given by the number of ways of obtaining this distribution. Note that the permutation of particles in the same energy level does not result in a different distribution. Hence, the number of ways of obtaining a given distribution is equal to $N!$ divided by the product of the number of permutations of particles within each energy level; or,

$$\mathscr{W} = \frac{N!}{N_1!N_2!N_3! \dots}. \tag{28}$$

In deriving Eq. (28) it has been assumed that the particles are identical but distinguishable, for example, by their positions— as in a crystal lattice. However, the molecules in a gas are indistinguishable, as well as identical. Thus, the probability given by Eq. (28) must be divided by $N!$. The resulting probability distribution for a gas is then given by

$$\mathscr{W} = \frac{1}{\displaystyle\prod_i N_i!}. \tag{29}$$

In addition, $\sum_i N_i = N$ and $\sum_i N_i \varepsilon_i = E$, because both the total number of particles and the total energy of the system must be conserved.

The logarithm of Eq. (29) yields

$$\ln \mathscr{W} = -\sum_i \ln N_i!, \tag{30}$$

and the most probable distribution is obtained by setting the differential of Eq. (30) equal to zero, namely

$$\delta \ln \mathscr{W} = 0 = -\delta \sum_i \ln N_i!. \tag{31}$$

When N_i is large, Stirling's approximation, $N_i! \approx N_i \ln N_i - N_i$, can be applied and Eq. (31) becomes

$$\delta \sum_i N_i \ln N_i - \delta \sum_i N_i = 0. \tag{32}$$

Three conditions are thus imposed to maximize the probability distribution, namely

(i)
$$\sum_i \ln N_i \, \delta N_i = 0, \tag{33}$$

which is the result of the development of Eq. (32),

(ii)
$$\sum_i \delta N_i = 0 \tag{34}$$

and

(iii)
$$\sum_i \varepsilon_i \, \delta N_i = 0. \tag{35}$$

Conditions (ii) and (iii) assure the conservation of the number of particles and the total energy, respectively.

The simultaneous imposition of the above three conditions can be achieved with the use of Lagrange's undetermined multipliers. In this method, two parameters α and β are introduced, which are determined from two additional conditions, as will be shown later. The general relation for the maximum probability distribution is then

$$\sum_i (\alpha + \beta \varepsilon_i + \ln N_i) \delta N_i = 0 \tag{36}$$

and

$$\alpha + \beta \varepsilon_i + \ln N_i = 0, \tag{37}$$

for each value of i. Then, the number of particles that occupy energy level i is given by

$$N_i = e^{-\alpha} e^{-\beta \varepsilon_i}. \tag{38}$$

However, in the case of degeneracies, more than one particle may occupy level i. If the degree of degeneracy is equal to g_i, then the distribution law of Eq. (38) takes the more general form

$$N_i = g_i e^{-\alpha} e^{-\beta \varepsilon_i}. \tag{39}$$

In statistical mechanics, the degree of degeneracy, g_i, is referred to as the statistical weight of level i.

From Eq. (39) the fraction of molecules in level i is given by

$$\frac{N_i}{N} = \frac{g_i e^{-\beta \varepsilon_i}}{\sum_i g_i e^{-\beta \varepsilon_i}}. \tag{40}$$

The denominator of Eq. (40), $\sum_i g_i e^{-\beta \varepsilon_i} = \mathscr{Z}$, is known as the partition function (*Fonction de partition*, in French) or sum-over-states (*Zustandssumme*, in German). As will be shown later, $\beta = 1/kT$, where T is the absolute temperature defined above and k is the Boltzmann constant.* Thus, the partition function for a given system is a function of the absolute temperature.

---

*In classical thermodynamics the (inexact) differential change in heat, δq, is related to the (exact) differential change in entropy, $\mathrm{d}S$, by

$$\mathrm{d}S = k\beta \, \delta q \equiv T^{-1} \, \delta q.$$

Thus, $k\beta$ (or T^{-1}) is an integrating factor for the heat change.

The total energy of the system is given by

$$E = \sum_i N_i \varepsilon_i, \tag{41}$$

$$= \frac{N}{\mathscr{Z}} \sum_i \varepsilon_i g_i \, \mathrm{e}^{-\varepsilon_i / kT}, \tag{42}$$

$$= \frac{NkT^2}{\mathscr{Z}} \frac{\partial \mathscr{Z}}{\partial T}. \tag{43}$$

The energy per mole of gas is then equal to

$$\tilde{E} = RT^2 \frac{\partial \ln \mathscr{Z}}{\partial T} \tag{44}$$

and the heat capacity per mole is given by

$$\tilde{C}_V = \left(\frac{\partial \tilde{E}}{\partial T} \right)_V = R \frac{\partial}{\partial T} \left(T^2 \frac{\partial \ln \mathscr{Z}}{\partial T} \right). \tag{45}$$

From Eqs (44) and (45) it is apparent that calculation of the energy and heat capacity of a system depends on the evaluation of the partition function as a function of temperature. In the general case of molecules with an internal structure, the energy distributions of the various degrees of freedom must be determined. This problem will be summarized in Chapter 4.

Chapter 3

The Distribution of Molecular Speeds

With the definition of the partition function $\mathscr{Z}$ and the identification $\beta = 1/kT$, Eq. (40) becomes

$$\frac{N_i}{N} = \frac{g_i\,\mathrm{e}^{-\varepsilon_i/kT}}{\mathscr{Z}}, \tag{46}$$

which expresses the fraction of molecules in level i. Equation (46) was obtained in the previous section with the assumption that the system was quantized. The application of this result in classical mechanics is appropriate in the limit that the separation between successive energy levels approaches zero. The energy of the system then becomes continuous.

The simple model of the kinetic theory of structureless particles presented in Chapter 1 was classical. The energy resulting from the displacement of a given molecule in, say, the x direction was given by $\frac{1}{2}m\dot{x}^2$. In this case, Eq. (46) can be applied assuming a continuous distribution of non-degenerate energy levels. Then, the number of molecules $\mathrm{d}N$ with velocities between $\dot{x}$ and $\dot{x} + \mathrm{d}\dot{x}$ is given by

$$\frac{\mathrm{d}N}{N} = \mathscr{A}\,\mathrm{e}^{-m\dot{x}^2/2kT}\,\mathrm{d}\dot{x}, \tag{47}$$

where $\mathscr{A}$ is a normalization constant. Thus,

$$\mathscr{A}\int_{-\infty}^{\infty} \mathrm{e}^{-m\dot{x}^2/kT}\,\mathrm{d}\dot{x} = 1, \tag{48}$$

which leads to $\mathscr{A} = \sqrt{m/2\pi kT}$. The resulting expression,

$$\frac{\mathrm{d}N}{N} = \sqrt{m/2\pi kT}\,\mathrm{e}^{-m\dot{x}^2/2kT}\,\mathrm{d}\dot{x}, \tag{49}$$

can easily be generalized to three dimensions in the form

$$\frac{dN}{N} = (m/2\pi kT)^{3/2}\, e^{-mu^2/2kT}\, d\dot{x}\, d\dot{y}\, d\dot{z}. \tag{50}$$

It is apparent from Fig. 4 that in spherical coordinates the volume element in Eq. (50) becomes equal to $u^2 \sin\theta\, du\, d\theta\, d\phi$. The integration over the angles θ and ϕ can be carried out to yield the factor 4π steradians; then

$$\frac{dN}{N} = 4\pi(m/2\pi kT)^{3/2}\, e^{-mu^2/2kT}\, u^2\, du. \tag{51}$$

Equation (51) is one form of the Maxwell–Boltzmann distribution law. As an example, this distribution is shown in Fig. 5 for nitrogen at various temperatures.

A number of useful relations can be obtained from Eq. (51). For example, the mean-square molecular speed can be expressed by

$$\overline{u^2} = \frac{1}{N}\int u^2\, dN, \tag{52}$$

$$= 4\pi(m/2\pi kT)^{3/2}\int_0^\infty e^{-mu^2/2kT}\, u^4\, du, \tag{53}$$

$$= \frac{3kT}{m}.^* \tag{54}$$

The corresponding translational energy is then given by

$$E = \tfrac{1}{2}Nm\overline{u^2} = \frac{3NkT}{2} = \tfrac{3}{2}n_0 RT, \tag{55}$$

which is identical to Eq. (20). This result justifies the relation $\beta = 1/kT$ suggested above. For the root-mean-square speed, Eq. (54) then yields

$$\sqrt{\overline{u^2}} = \sqrt{\frac{3kT}{m}} = \sqrt{\frac{3RT}{M}}, \tag{56}$$

where M is the mass of one mole of particles.

Another useful quantity is the average molecular speed. It can be defined by

$$\overline{u} = \frac{1}{N}\int u\, dN. \tag{57}$$

---

*The integral in Eq. (53) can be found from the general relation

$$\int_0^\infty x^{2n}\, e^{-ax^2}\, dx = \frac{(2n-1)!!}{2(2a)^n}\sqrt{\frac{\pi}{a}},$$

where n is a positive integer and $a > 0$.

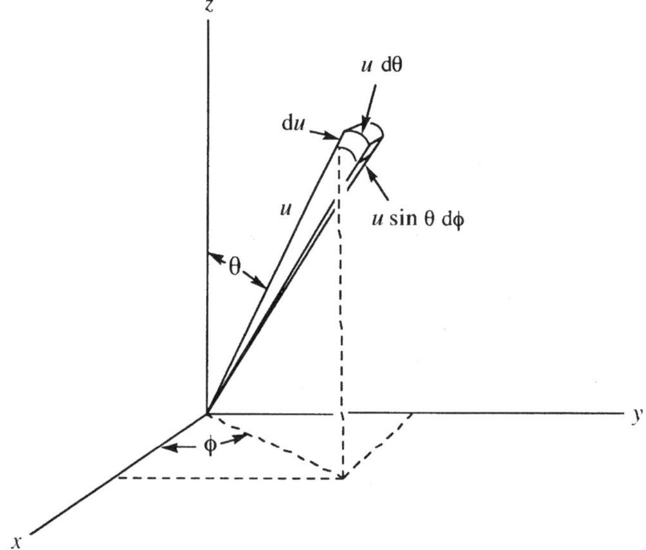

Fig. 4 Definition of the spherical coordinate system

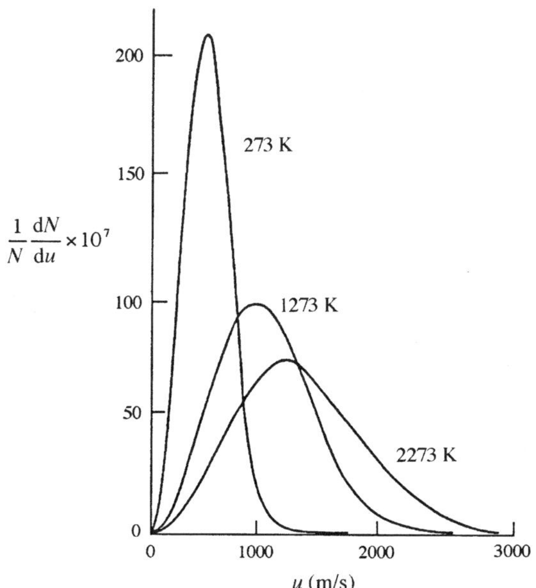

Fig. 5 Distribution of molecular speeds in nitrogen

Substitution of Eq. (51) in Eq. (57) yields

$$\bar{u} = 4\pi (m/2\pi kT)^{3/2} \int_0^\infty e^{-mu^2/2kT} u^3 \, du, \qquad (58)$$

$$= \sqrt{8kT/\pi m} = \sqrt{8RT/\pi M}. \qquad (59)$$

In certain applications it is necessary to calculate the most probable molecular speed. This quantity corresponds to the position of the maximum in the Maxwell–Boltzmann distribution given by Eq. (51). By setting the derivative equal to zero, namely

$$\frac{d}{du}\left(\frac{dN}{N}\right) = 4\pi (m/2\pi kT)^{3/2}\left(2u - \frac{mu^3}{kT}\right)e^{-mu^2/2kT} = 0, \qquad (60)$$

it is found that the most probable molecular speed is given by

$$u_{\max} = \sqrt{2kT/m} = \sqrt{2RT/M}. \qquad (61)$$

The various speeds, $u_{\max}$, $\bar{u}$ and $\sqrt{\overline{u^2}}$ are indicated on the distribution curve shown in Fig. 6. They are related, since from Eqs (56), (59) and (61), $\bar{u} = 1.128 u_{\max}$ and $\sqrt{\overline{u^2}} = 1.225 u_{\max}$.

In the simple kinetic theory model introduced in Chapter 1, the particles were treated as point masses. Thus, the actual volume occupied by the molecules was neglected. However, in most applications it is necessary to attribute a certain

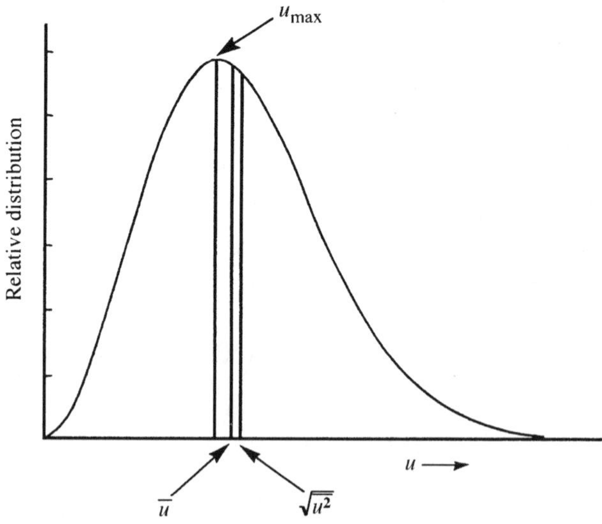

Fig. 6 Comparison of $u_{\max}$, $\bar{u}$ and $\sqrt{\overline{u^2}}$; the curve represents the Maxwell–Boltzmann distribution

volume to each molecule. In the following development it will be assumed that each of the identical molecules in the gas is spherical, with a diameter σ. When the chosen molecule a, moves among the others it sweeps out a cylindrical volume of diameter 2σ, as shown in Fig. 7(a). On average, the length of the cylinder is equal to $\bar{u}$ per second, corresponding to a volume of $\pi\sigma^2\bar{u}$. Hence, if there are n molecules (at rest) per unit volume of gas, molecule a undergoes $n\pi\sigma^2\bar{u}$ collisions per second with the surrounding ones. A factor of $\sqrt{2}$ is usually introduced to account for the movement of the other molecules. Thus, it is the relative velocities of two colliding molecules that must be considered (see Fig. 7(b)). It should be noted that the factor $\sqrt{2}$ is approximate, although quite adequate for practical purposes. The number of collisions per second of a given molecule (a) with all other molecules is then given by

$$Z_a = \sqrt{2}\pi n\sigma^2\bar{u}, \tag{62}$$

and the number of binary collisions per second between like molecules becomes equal to

$$Z_{aa} = \frac{Z_a n}{2} = 2n^2\sigma^2\sqrt{\frac{\pi kT}{m}} = 2n^2\sigma^2\sqrt{\frac{\pi RT}{M}}, \tag{63}$$

where the expression has been divided by two to avoid counting each collision twice. In the case of gas mixtures, Eq. (63) must be modified to obtain the frequency of collisions between unlike molecules [see Part II, Eq. (16)].

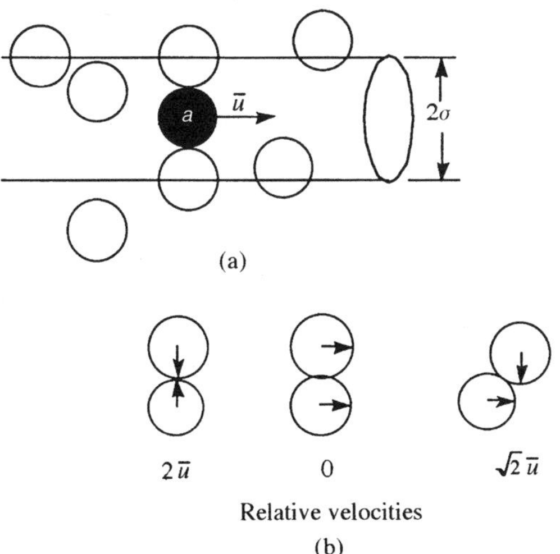

Fig. 7 Collisions of molecules of diameter σ: (a) The chosen molecule a, collides with all stationary molecules within the cylinder of diameter 2σ. (b) The relative velocities in a binary collision can vary from 0 to $2\bar{u}$ for two molecules of average velocity $\bar{u}$

Equation (62) allows a definition to be made of the mean free path of a given molecule. It is introduced here in the form

$$\lambda_0 = \frac{\overline{u}}{Z_a} = \frac{1}{\sqrt{2}\pi n \sigma^2}. \tag{64}$$

This quantity can be interpreted as the average distance traversed by a given molecule between two successive collisions with its neighbors. It will be employed in Chapter 5 in the analysis of transport phenomena—viscosity, thermal conductivity and diffusion.

Chapter 4

Molecular Energies

As everybody knows, molecules are composed of electrons and nuclei. Even for hydrogen, in which the nucleus is a single proton, an electron is approximately 1800 times lighter. Thus, the electrons move very rapidly compared with the nuclei, forming the so-called electron cloud during a period of time in which the nuclei undergo only slight displacements. This physical argument is the basis of the Born–Oppenheimer approximation. It permits the total energy of a molecule to be separated into two parts, namely

$$\varepsilon_{\text{molecule}} \approx \varepsilon_{\text{electrons}} + \varepsilon_{\text{nuclei}}. \tag{65}$$

This approximation is an excellent one, although a few spectroscopic effects arise from the coupling between electronic and nuclear motion.

The energy associated with the nuclear motion can be separated into an external contribution (overall translation of the molecule in space) and an internal part. The latter consists of the rotation of the molecule and the molecular vibrations. The separation of internal and external energies is valid for molecules at low pressures, where intermolecular forces are weak. In this case, the molecule rotates about its center of gravity.

The separation of the internal molecular energy into rotational and vibrational contributions is, in general, a poorer approximation, as the Coriolis forces can couple these motions. However, this separation will be assumed to be valid in the present analysis of the molecular degrees of freedom, and the energy of nuclear motion will be written in the form

$$\varepsilon_{\text{nuclei}} \approx \varepsilon_{\text{translation}} + \varepsilon_{\text{rotation}} + \varepsilon_{\text{vibration}}. \tag{66}$$

In the following development of the expressions for the partition functions for the various degrees of molecular freedom, the molecules will be assumed to remain in the ground electronic state. At moderate temperatures this limitation is appropriate. In this case, the origin of the energy scale can be defined by $\varepsilon_{\text{electrons}} = 0$, which leads to the electronic partition function $\mathscr{Z}_{\text{electrons}} = 1$, if the ground state is non-degenerate. However, at very high temperatures the population of excited electronic states cannot be neglected (see Chapter 16).

4.1   TRANSLATION

The derivation of the expression for the translational partition function requires an elementary quantum-mechanical treatment. For this purpose, it can be assumed that m, the total mass of the molecule, is located at the center of gravity. Its displacement within a box of dimensions a, b, c, as employed in Chapter 1 is equivalent to that of a point mass m. This problem, which is discussed in all elementary books on quantum mechanics, is known as 'the particle in a box'.

The problem in one dimension, say x, is formulated with the aid of Schrödinger's equation, which is $\hat{H}\psi = \varepsilon_t\psi$. Here, $\hat{H}$ is Hamilton's operator, which corresponds to the energy. In this application it is that which represents the translational energy in one dimension, while ε_t is the corresponding value of the energy. The equation for the translational wave function is then[*]

$$-\frac{h^2}{8\pi^2 m}\frac{\mathrm{d}^2\psi_t}{\mathrm{d}x^2} + \mathscr{V}(x)\psi_t = \varepsilon_t\psi_t. \tag{67}$$

The potential energy, which is represented here by $\mathscr{V}(x)$, is given by

$$\mathscr{V}(x) = \begin{cases} 0, & 0 < x < a \\ \infty, & x = 0,\, a \end{cases}. \tag{68}$$

This function corresponds to the assumptions made in the classical treatment of Chapter 1, namely the particle is not subjected to any forces within the box, but suffers inelastic collisions at each wall. Inside the box, where $\mathscr{V}(x) = 0$, the solutions to Eq. (67) take the form

$$\psi_t = B\sin\left[\left(\frac{2\pi}{h}\sqrt{2m\varepsilon_t}\right)x + \gamma\right], \tag{69}$$

which describes the deBroglie wave of the free particle in the x direction. The amplitude of the wave, B, is usually chosen to normalize the wave function, while the phase, γ, is determined by a boundary condition. The wavelength is given by $\lambda = h/(2m\varepsilon_t)^{1/2}$, and, since

$$\varepsilon_t = \tfrac{1}{2}m\dot{x}^2 = \frac{p^2}{2m}, \tag{70}$$

$$\lambda = \frac{h}{p}. \tag{71}$$

Equation (71), which was first proposed by deBroglie, expresses the relation between the wavelength of the descriptive wave and the momentum p of the particle. Although it was derived here in one dimension, Eq. (71) is general for three-dimensional motion.

---

[*]Here, the kinetic energy is given by $\hat{p}_x^2/2m$, where $\hat{p}_x = (\hbar/i)(\partial/\partial x)$ is the momentum operator and $\hbar \equiv h/2\pi$.

In the one-dimensional motion of the particle it strikes the walls of the box at $x = 0$ and $x = a$, where from Eq. (68) the potential function becomes infinite. Accordingly, an acceptable wave function must vanish at these two points. The imposition of these boundary conditions requires that $\gamma = 0$ and that the coefficient of x in Eq. (69) be equal to $\pi N/a$, where N is a positive integer. The resulting wave function becomes

$$\psi_N = \frac{1}{\sqrt{2\pi}} \sin \frac{N\pi x}{a}, \quad N = 1, 2, 3, \ldots, \tag{72}$$

and the energy is given by

$$\varepsilon_x = \frac{h^2 N^2}{8ma^2}, \tag{73}$$

with N a positive integer, as above. The translational partition function for the degree of freedom in the x direction can then be written in the form

$$\mathscr{Z}^{(x)} = \sum_N e^{-N^2 h^2/(8ma^2 kT)} \tag{74}$$

in this non-degenerate system ($g_N = 1$).

The results obtained here for the particle in a one-dimensional box are summarized in Figs 8 and 9. The wave functions given by Eq. (72) are plotted

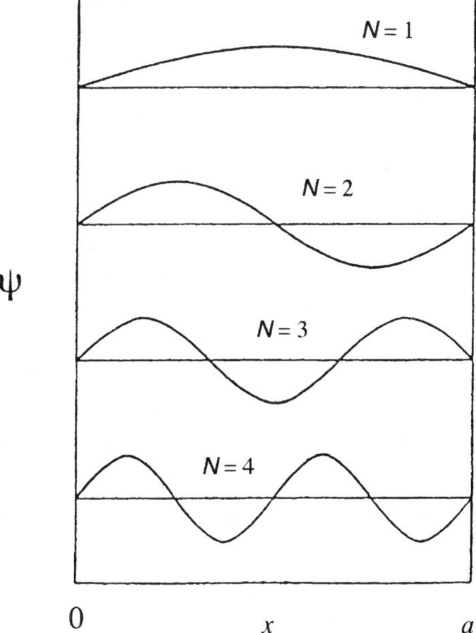

Fig. 8 Particle-in-a-box wave functions for different values of the quantum number N

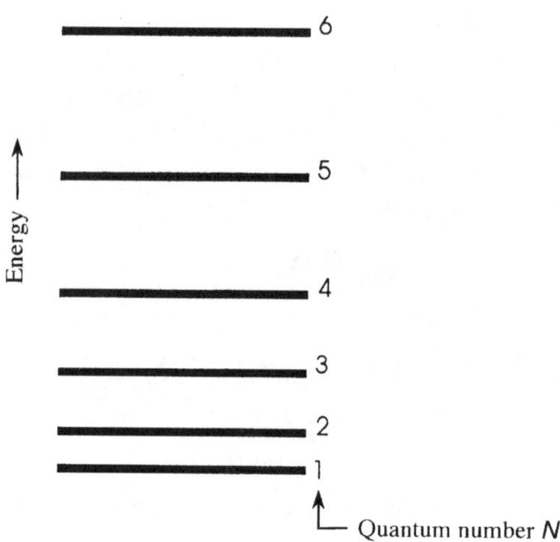

Fig. 9   Energy levels for a particle in a one-dimensional box

in Fig. 8, where the analogy with the vibrational modes of a string with fixed ends should be obvious. More important in the present application is Eq. (73), which yields the energy levels shown in Fig. 9. It is evident from Eq. (73) that the spacing between successive levels increases quadratically with N. Furthermore, the separation between levels decreases as the quantity ma^2 increases. As the dimension a is very large on a molecular scale, for virtually all molecules, the energy distribution given by Eq. (73) approaches a continuum. In other words, the deBroglie wave length of the particle is very short compared with a. At moderate temperatures almost all of the molecules are to be found in the lower levels. Thus, to a good approximation, the sum in Eq. (74) can be replaced by an integral over the energy levels, namely

$$\mathscr{Z}^{(x)} \approx \int_0^\infty e^{-N^2 h^2/(8ma^2 kT)} \, dN. \tag{75}$$

$$= \frac{a\sqrt{2\pi mkT}}{h}. \tag{76}$$

Obviously, the above derivation can be repeated for the other two Cartesian directions. Since the energies are additive, the partition function for the three-dimensional translation of the particle can be written as a product, namely

$$\mathscr{Z}_{\text{translation}} = \mathscr{Z}^{(x)} \, \mathscr{Z}^{(y)} \, \mathscr{Z}^{(z)} \tag{77}$$

$$= \frac{abc}{h^3}(2\pi mkT)^{3/2} = \frac{V}{h^3}(2\pi mkT)^{3/2}, \tag{78}$$

where V is the volume of the box. Equation (78) leads to

$$\ln \mathscr{Z}_{\text{translation}} = \tfrac{3}{2} \ln T + \text{constant} \tag{79}$$

and

$$\frac{\partial \ln \mathscr{Z}_{\text{translation}}}{\partial T} = \frac{3}{2T}. \tag{80}$$

The translational energy of the system is then given by

$$E_{\text{translation}} = \tfrac{3}{2} n_{\text{o}} RT \tag{81}$$

and the energy per mole by

$$\tilde{E}_{\text{translation}} = \tfrac{3}{2} RT. \tag{82}$$

Finally, the translational contribution to the heat capacity at constant volume, per mole of gas, is then given by

$$\hat{C}_{V(\text{translation})} = \tfrac{3}{2} R, \tag{83}$$

that is, $\tfrac{1}{2} R$ for each of the three degrees of translational freedom. Note that this result is identical to that obtained from elementary kinetic theory [see Eq. (23)].

The Schrödinger equation for the internal degrees of freedom of an isolated molecule can be separated into the rotational and vibrational parts; however, several assumptions must be made. The Coriolis forces are neglected and, furthermore, the amplitudes of the vibrational modes are considered to be infinitesimal. The latter assumption leads to the rigid-rotor, harmonic-oscillator approximation, which is usually employed in elementary treatments of this problem.[*]

4.2 ROTATION

The form of the Schrödinger equation for the molecular rotation depends on the type of rotor; that is, the geometry of the molecule in its equilibrium configuration. Four types of rotor can be distinguished, depending on the moments of inertia I_a, I_b, and I_c, with respect to the principal axes a, b, c. The rotors are defined as:

(i) Linear: $I_a = I_b$, $I_c = 0$.

In this case the rotational energy is given by $\varepsilon_{\text{rotation}} = (h^2/8\pi^2 I)J(J+1)$, where the quantum number $J = 0, 1, 2, \ldots$ and $I = I_a = I_b$. For readers who are familiar with the quantum-mechanical treatment of the hydrogen atom, it should be noted that the quantum number J introduced here is analogous to the

---

[*]The harmonic approximation is specified for a diatomic molecule by Eq. (95).

azimuthal quantum number ℓ. In the present case as well, the energy levels are of degeneracy, $g_J = 2J + 1$, as the quantum number M (analogous to the magnetic quantum number m for the H atom) can take on the values, $M = 0$, ± 1, ± 2, ..., $\pm J$.

(ii) Spherical: $I_a = I_b = I_c$.

Here, the energy is again given by $(h^2/8\pi^2 I)J(J+1)$, with $I_a = I_b = I_c = I$, although the degeneracy is now equal to $(2J+1)^2$.

(iii) Symmetric: $I_a = I_b \neq I_c \neq 0$,

where principal axis c is taken to be the axis of symmetry of the molecule—for example, the three-fold axis of the ammonia molecule. The expression for the rotational energy of this rotor depends on two independent quantum numbers, J and K. Thus,

$$\varepsilon_r = \frac{h^2}{8\pi^2 I_a} J(J+1) + \frac{K^2 h^2}{8\pi^2}\left(\frac{1}{I_c} - \frac{1}{I_a}\right), \tag{84}$$

where $K = 0$, ± 1, ± 2, ..., $\pm J$. It should be noted that if $I_c > I_a = I_b$, the second term in Eq. (84) is negative; the rotor is then said to be an oblate spheroid, as is the earth, which is slightly flattened at the poles (a pancake represents the extreme case). However, $I_c < I_a = I_b$, defines a prolate rotator, as is an American football. The spectroscopic consequences of these two situations are quite different, as the two types of rotor are easily distinguished by their rotational spectra.

(iv) Asymmetric: $I_a \neq I_b \neq I_c$.

This case is the most general one. It is also the most difficult, as no analytical expression can be derived for the energy levels. The reader is referred to advanced books on spectroscopy for a discussion of this problem.

For linear molecules and, hence, diatomics, the rotational partition function can be written as

$$\mathscr{Z}_{\text{rotation}} = \sum_J (2J+1)\,e^{-h^2 J(J+1)/8\pi^2 IkT}. \tag{85}$$

The sum appearing in Eq. (85) would be expected to extend over all values of J. However, in the case of molecules with a center of symmetry the nuclear spins must be considered. For example, molecular hydrogen exists in two forms, *ortho* and *para*, the sum being over odd or even values of J, respectively. This particular application is developed further in Chapter 11.

Returning now to linear molecules lacking a center of symmetry, Eq. (85) is applicable with $J = 0, 1, 2, 3, \ldots, \infty$. It is often useful to define a rotational temperature by the relation $\theta_{\text{rotation}} = h^2/8\pi^2 Ik$. At temperatures significantly greater than this value, the sum in Eq. (85) can be replaced by an integral over J.

This operation is equivalent to the passage from the quantum-mechanical to the classical description of the rotational motion of the molecule. It yields

$$\mathscr{Z}_{\text{rotation}} \approx \frac{8\pi^2 IkT}{h^2} = \frac{T}{\theta_{\text{rotation}}}, \tag{86}$$

for $T \gg \theta_{\text{rotation}}$. In this limit the definition of the partition function leads to the relation

$$\ln \mathscr{Z}_{\text{rotation}} = \ln T + \text{constant} \tag{87}$$

and

$$\frac{\partial \ln \mathscr{Z}_{\text{rotation}}}{\partial T} = \frac{1}{T}. \tag{88}$$

The rotational energy of linear molecules can then be written in the form

$$E_{\text{rotation}} = n_{\text{o}} RT^2 \left(\frac{1}{T}\right) = n_{\text{o}} RT, \tag{89}$$

or, for the molar quantities,

$$\tilde{E}_{\text{rotation}} = RT \tag{90}$$

and

$$\hat{C}_{V(\text{rotation})} = R. \tag{91}$$

This result corresponds to a contribution of $\frac{1}{2}R$ for each of the two degrees of rotational freedom of a linear molecule.

Non-linear polyatomic molecules require further consideration, depending on their classification under (ii), (iii) or (iv) above. Since three degrees of rotational freedom are now available, three (rather than two) quantum numbers are necessary to specify the rotational states. For spherical and symmetric rotors [(ii) and (iii)] analytical expressions for both the energies and the statistical weights can be obtained, yielding expressions for the rotational partition functions. In the case of asymmetric rotors, the Schrödinger equation for the molecular rotation cannot be solved analytically, although numerical values of the energy have been tabulated.

In the classical, high temperature limit, the rotational partition function for a non-linear molecule is given by

$$\mathscr{Z}_{\text{rotation}} \approx \frac{\sqrt{\pi}}{\sigma h^3} \sqrt{I_a I_b I_c} (8\pi^2 kT)^{3/2}, \tag{92}$$

where $I_a = I_b = I_c$ for spherical rotors and $I_a = I_b$ for symmetric rotors. The symmetry number σ depends on the structure of the molecule. For example, a molecule such as H_2O, which belongs to point group C_{2v}, has a two-fold axis of symmetry, leading to $\sigma = 2$. For ammonia (C_{3v}), $\sigma = 3$, while for methane (T_d)

and benzene (C_{6v}), $\sigma = 4 \times 3 = 12$ (4 three-fold axes) and $\sigma = 6 \times 2 = 12$ (6 two-fold axes), respectively. In all cases, Eq. (92) yields

$$\hat{C}_{V\text{(rotation)}} = \tfrac{3}{2}R \tag{93}$$

or $\tfrac{1}{2}R$ for each of the three degrees of rotational freedom of non-linear molecules.*

4.3 VIBRATION

Thus far, the three translational degrees of freedom of a molecule have been considered and the contributions to the heat capacity have been evaluated. As for the rotations, linear molecules have been shown to have two degrees of freedom, while non-linear polyatomic molecules have three. The remaining degrees of freedom of a molecule are associated with the vibrations. As there are three degrees of freedom for each of the $\mathcal{N}$ atoms in a molecule, $3\mathcal{N} - 5$ remain for the vibrations of a linear molecule, or $3\mathcal{N} - 6$ for a non-linear one.

The evaluation of the vibrational partition function is summarized here for a diatomic molecule, which has but one degree of vibrational freedom. This result can be generalized for each normal mode of vibration of a polyatomic molecule.

For a diatomic molecule, the intramolecular potential function, which has the general form shown in Fig. 10, determines the force between the atoms. It is, therefore, characteristic of the chemical bond. The position of the minimum

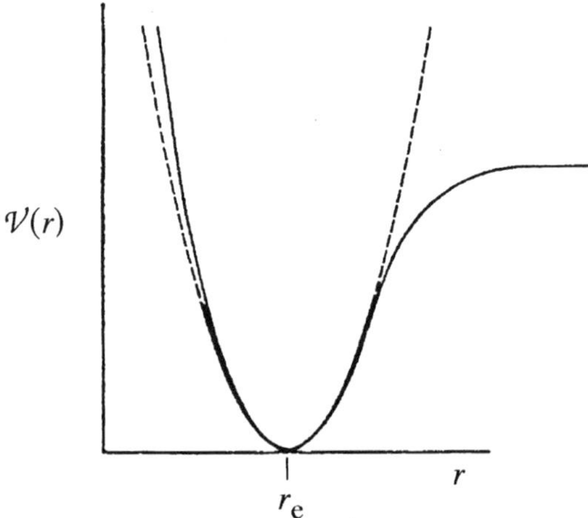

Fig. 10 Potential function for a diatomic molecule; the dotted curve represents the harmonic approximation

---

*Here, again, it has been assumed that T is significantly greater than the rotational temperature.

corresponds to the equilibrium bond length r_e, while a value of the dissociation energy is reached as the interatomic distance r approaches infinity.

It is convenient to develop the potential function of Fig. 10 in a Taylor series about the equilibrium distance r_e. Thus,

$$\mathscr{V}(r) = \mathscr{V}_0 + \left(\frac{d\mathscr{V}}{dr}\right)_0 (r - r_e) + \tfrac{1}{2}\left(\frac{d^2\mathscr{V}}{dr^2}\right)_0 (r - r_e)^2 + \dots . \tag{94}$$

The first term on the right-hand side of Eq. (94) can be set equal to zero as a definition of the origin of the potential energy scale. The second term is also equal to zero, as the tangent is horizontal at $r = r_e$. The first non-vanishing term contains the second derivative, the force constant $k \equiv (d^2\mathscr{V}/dr^2)_0$, which is also characteristic of the chemical bond. Higher terms in Eq. (94) are responsible for the anharmonicity of the vibration. They are neglected in the harmonic approximation, which allows the potential function to be written in the form

$$\mathscr{V}(r) = \tfrac{1}{2}k(r - r_e)^2, \tag{95}$$

as shown by the dotted curve in Fig. 10. The Schrödinger equation for the vibrational wave function is then

$$-\frac{\hbar^2}{2\mu}\left(\frac{d^2\psi_V}{d\xi^2}\right) + \tfrac{1}{2}k\xi^2\psi_V = \varepsilon_V\psi_V, \tag{96}$$

where $\xi = r - r_e$ and μ is the reduced mass of the two atoms of masses m_a and m_b. The reduced mass is defined by

$$\frac{1}{\mu} = \frac{1}{m_a} + \frac{1}{m_b}. \tag{97}$$

In Eq. (96) the wavefunction ψ_V defines the vibrational state of the molecule, with energy ε_V, where $V = 0, 1, 2, \dots$, is the vibrational quantum number.

The solution of Eq. (96) leads to expressions for the wave function in terms of Hermite polynomials and energy. The latter is given by

$$\varepsilon_V = h\nu_0(V + \tfrac{1}{2}), \tag{98}$$

where

$$\nu_0 = \frac{1}{2\pi}\sqrt{\frac{k}{\mu}} \tag{99}$$

is the classical frequency of vibration. The energy levels determined by Eq. (98) are, in this case, non-degenerate, i.e. $g_V = 1$. The vibrational partition function can then be written as

$$\mathscr{Z}_{\text{vibration}} = e^{-h\nu_0/kT}\sum_V e^{-(h\nu_0/2kT)V} \tag{100}$$

and

$$\ln \mathscr{Z}_{\text{vibration}} = -\frac{h\nu_0}{2kT} + \ln \sum_v e^{-(h\nu_0/kT)V}. \tag{101}$$

The first term on the right-hand side of Eq. (101) yields an energy per mole of $N_0(\frac{1}{2}h\nu_0)$. This, so-called, zero-point energy exists even at the temperature of absolute zero, where molecules are in the fundamental level, $v = 0$. However, it makes no contribution to the heat capacity.

The second term on the right-hand side of Eq. (101) can be expanded in a series,

$$\sum_v e^{-(h\nu_0/kT)V} = \sum_v (e^{-h\nu_0/kT})^V \tag{102}$$

$$= \sum_V \zeta^V = 1 + \zeta + \zeta^2 + \ldots = \frac{1}{1-\zeta}, \tag{103}$$

where $\zeta = h\nu_0/kT$. Thus,

$$\sum_V e^{-(h\nu_0/kT)V} = (1 - e^{-h\nu_0/kT})^{-1} \tag{104}$$

and

$$\ln \mathscr{Z}_{\text{vibration}} = -\frac{h\nu_0}{2kT} - \ln(1 - e^{-h\nu_0/kT}). \tag{105}$$

The vibrational energy per mole of a diatomic gas is then given by

$$\tilde{E}_{\text{vibration}} = N_0 kT^2 \frac{\partial \ln \mathscr{Z}_{\text{vibration}}}{\partial T} \tag{106}$$

$$= N_0(\tfrac{1}{2}h\nu_0) - \frac{N_0 h\nu_0\, e^{-h\nu_0/kT}}{e^{-h\nu_0/kT} - 1} \tag{107}$$

and the heat capacity per mole becomes equal to

$$\tilde{C}_{V(\text{vibration})} = N_0 h\nu_0 \frac{\partial}{\partial T}\frac{1}{e^{h\nu_0/kT} - 1} \tag{108}$$

$$= R\left(\frac{h\nu_0}{kT}\right)^2 \frac{e^{h\nu_0/kT}}{(e^{h\nu_0/kT} - 1)^2}. \tag{109}$$

In the high temperature limit ($kT \gg h\nu_0$), Eq. (109) reduces to $\tilde{C}_{V(\text{vibration})} \approx R$. However, this condition is fulfilled only for molecules composed of two heavy atoms, or at relatively high temperatures.

The results obtained above for a diatomic molecule can be generalized for polyatomic molecules. As pointed out earlier, non-linear polyatomic molecules composed of $\mathscr{N}$ atoms have $3\mathscr{N} - 6$ degrees of vibrational freedom. These vibrations can be represented in the harmonic approximation as an ensemble of normal modes, each of which has a characteristic frequency. The quantum-mechanical treatment of this problem leads to expressions for the wave functions and energies analogous to those for the diatomic molecule. Thus, the vibrational wave function for a non-linear polyatomic molecule can be written in the form

$$\psi_{\text{vibration}} = \prod_k \psi_k(v_k), \tag{110}$$

where the product extends over the $3\mathcal{N}' - 6$ normal modes, or $3\mathcal{N}' - 5$ if the molecule is linear. The index k identifies the normal mode for which v_k is the vibrational quantum number. The corresponding expression for the energy is

$$\varepsilon_{\text{vibration}} = \sum_k h\nu_k(v_k + \tfrac{1}{2}). \tag{111}$$

The sum in Eq. (111) is over the $3\mathcal{N}' - 6$ (or $3\mathcal{N}' - 5$) normal modes of the molecule. It should be noted, however, that degeneracies often arise, depending on the symmetry of the molecule. In these cases, there may be more than one wave function for a given energy level. The reader is referred to books on vibrational spectroscopy for the application of the theory of groups to the analysis of molecular symmetry and the determination of degeneracies, as well as optical selection rules.

The evaluation of the various contributions to the energy of a molecule can now be summarized, as given in Eq. (66),

$$\varepsilon_{\text{nuclei}} \approx \varepsilon_{\text{translation}} + \varepsilon_{\text{rotation}} + \varepsilon_{\text{vibration}}, \tag{66}$$

where the electronic energy has been chosen equal to zero for the system in the ground electronic state. With the aid of the partition function for the ensemble of molecules,

$$\mathscr{Z} = \mathscr{Z}_{\text{translation}} \cdot \mathscr{Z}_{\text{rotation}} \cdot \mathscr{Z}_{\text{vibration}} \tag{112}$$

and the heat capacity per mole, $\tilde{C}_V$, as well as all of the thermodynamic quantities, can be evaluated. For diatomic molecules, the heat capacity as a function of temperature is shown schematically in Fig. 11. The classical, high-temperature limits for both linear and non-linear molecules are given in Table 2.

Table 2   Classical contributions to the molar heat capacity

Number of degrees of freedom			Maximum contribution to $\tilde{C}_V$
Linear molecules:			
$3\mathcal{N}' =$	3	translation	$\frac{3}{2}R$
	2	rotation	$\frac{2}{2}R$
	$3\mathcal{N}' - 5$	vibration	$(3\mathcal{N}' - 5)R$
Non-linear molecules:			
$3\mathcal{N}' =$	3	translation	$\frac{3}{2}R$
	3	rotation	$\frac{3}{2}R$
	$3\mathcal{N}' - 6$	vibration	$(3\mathcal{N}' - 6)R$

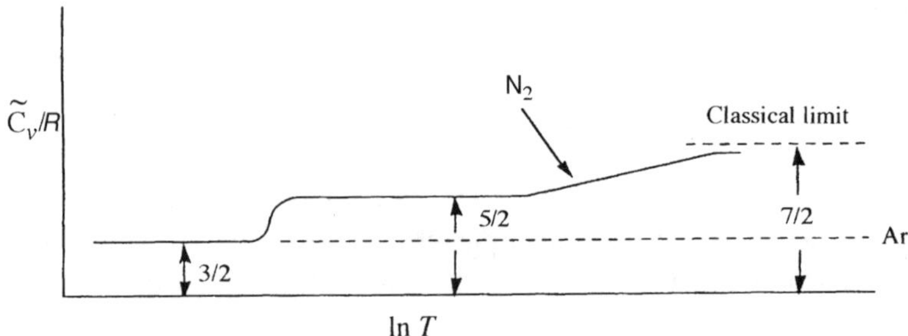

Fig. 11   Heat capacity ($\tilde{C}_V/R$) as a function of temperature

It should be emphasized that all of these results depend on the assumption of thermal equilibrium among the various degrees of freedom of the molecules. When the system is subjected to a relatively rapid external perturbation, such as a shock wave or an absorbed pulse of electromagnetic radiation, the re-equilibration of the system will be obtained as a result of various relaxation processes. Some examples of these phenomena will be treated as applications in Chapter 13.

Chapter 5

Transport Phenomena

The preceding chapters have been devoted to the simple kinetic theory of a gas that, from the macroscopic point of view, is in complete equilibrium. The present chapter, however, will be concerned with phenomena in a gas that is usually in a steady state, although not, strictly speaking, in equilibrium. Such a gas is often referred to as non-uniform. In the following analyses, the disturbances that produce local departures from equilibrium are assumed to be neither so great nor so rapid as to upset the Maxwell–Boltzmann distribution. Furthermore, since they are based on the kinetic theory of ideal gases, the conclusions reached become invalid with increasing pressure.

The physical phenomena involved in a non-uniform gas are called the transport properties, e.g. viscosity, thermal conductivity and diffusion. In an ionized gas, an additional phenomenon is involved, the electrical conductivity (see Chapter 16). Very simple models of these properties will now be presented, as they lead to a basic understanding of, although not necessarily to good agreement with, experimental measurements.

5.1 VISCOSITY

An amusing analogy has been made by Moore, which should aid in understanding the basis physics of gas viscosity.[*]

Two railroad trains are moving in the same direction, but at different speeds, on parallel tracks. The passengers on these trains amuse themselves by jumping back and forth from one to the other. When a passenger jumps from the more rapidly moving train to the slower one he transports momentum of amount mu, where m is his mass and u the velocity of his train. He tends to speed up the more slowly moving train when he lands upon it. A passenger who jumps from the slower to the faster train, on the other hand, tends to slow it down. The net result of the jumping game is thus a tendency to equalize the velocities of the two trains. An observer

---

[*]From *Physical Chemistry* by Walter J. Moore, quoted by permission of Prentice-Hall, Inc., Englewood Cliffs, New Jersey, USA.

from afar who could not see the jumpers might simply note this result as a frictional drag between the trains.

Consider now two layers of unit area within a gas, as represented in Fig. 12, which move at different velocities. Let the distance between the layers be equal to λ_o, the mean free path defined by Eq. (64). The difference in the velocities of the two layers in the y direction is due to the z component of the velocity gradient within the gas. Thus,

$$\Delta \dot{y} = \lambda_o \left(\frac{\partial \dot{y}}{\partial z} \right). \tag{113}$$

The viscosity of the gas results from the momentum transferred by molecules that pass from one layer to the other. Thus, if $(\partial \dot{y}/\partial z)$ is positive, then layer *a* moves faster than layer *b* and a molecule that passes from *a* to *b* will accelerate layer *b* and slow down layer *a* accordingly. Furthermore, a molecule moving in the opposite direction will have the same net effect. Each molecule of mass *m* then contributes a momentum exchange of $m\Delta \dot{y}$ when passing from one layer to the other. If there are *n* molecules per unit volume, then the average rate of change of momentum, and hence the average force because of the interacting layers, is given by

$$f = -2mn\bar{\dot{z}}\lambda_o \left(\frac{\partial \dot{y}}{\partial z} \right), \tag{114}$$

where $\bar{\dot{z}} = \sqrt{kT/2\pi m}$ is the average speed of molecules moving parallel to the z axis. The expression for $\bar{\dot{z}}$ can be obtained from Eq. (49) with x replaced by z. Note that the minus sign in Eq. (114) indicates that the force or 'drag' tends to counter the relative motions of the two layers.

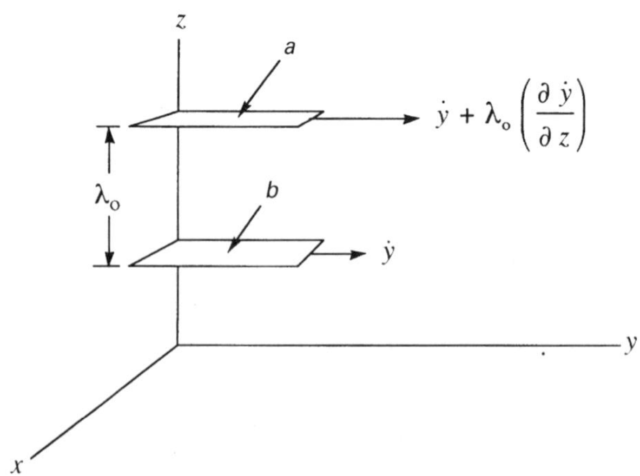

Fig. 12   Model for the determination of the viscosity of a gas

Following Newton, the force between the two layers can be written in the form

$$f = -\eta \left(\frac{\partial \dot{y}}{\partial z} \right), \tag{115}$$

where η is the viscosity coefficient—or, simply, the viscosity. Comparison of Eqs (114) and (115) leads to

$$\eta = 2mn\bar{\dot{z}}\lambda_{\mathrm{o}},$$

$$= \frac{1}{\pi\sigma^2} \sqrt{\frac{kTm}{\pi}}. \tag{116}$$

In terms of molar quantities, Eq. (116) can be written in the form

$$\eta = \frac{M}{\pi\sigma^2 N_{\mathrm{o}}} \sqrt{\frac{RT}{\pi M}}. \tag{117}$$

It can be concluded from this result that for an ideal gas composed of 'hard spheres' of diameter σ, the viscosity should be independent of the concentration and should vary as the square root of the absolute temperature. This result is in qualitative agreement with experimental results, although it is often necessary to introduce a numerical correction factor. It may be surprising that Eq. (116) does not contain *n*. However, it should be noted that in this simple model the effects of gas density on the separation between layers and the number of molecules passing between them cancel.

5.2   THERMAL CONDUCTIVITY

The thermal conductivity of a gas can be treated with the use of the same simple model employed above in the analysis of gas viscosity. The two layers *a* and *b* will now be assumed to be stationary, but at different uniform temperatures. A molecule passing from one layer to the other will transport an energy $c_V(\partial T/\partial z)$, where c_V is the heat capacity per molecule.

Therefore, by analogy with Eq. (114), the net rate of change of energy will be given by

$$\Delta\dot{\varepsilon} = -2n\bar{\dot{z}}\lambda_{\mathrm{o}}c_V \left(\frac{\partial T}{\partial z} \right). \tag{118}$$

Thermal conductivity can be defined as the quantity of heat transported across a unit area in unit time, because of a unit temperature gradient. Since heat always flows in the direction opposite to that of the temperature gradient,

$$\Delta\dot{\varepsilon} = -\kappa \left(\frac{\partial T}{\partial z} \right), \tag{119}$$

where the thermal conductivity κ is then positive. Comparison of Eqs (118) and (119) yields

$$\kappa = 2n\bar{\bar{z}}\lambda_{o}c_{V},$$

$$= \frac{\tilde{C}_{V}}{\pi\sigma^{2}N_{o}}\sqrt{\frac{RT}{\pi M}}. \tag{120}$$

It is of interest to compare the results obtained above for these two quite different physical phenomena. On the basis of this very simplified kinetic theory treatment, the thermal conductivity is given by Eq. (120) and the viscosity by Eq. (117). The ratio of these two expressions yields

$$\frac{\kappa}{\eta} = \frac{\tilde{C}_{V}}{M}, \tag{121}$$

or

$$\frac{\kappa M}{\eta\tilde{C}_{V}} = 1. \tag{122}$$

This relation is at least qualitatively verified by experiment, although the numerical value of the left-hand side of Eq. (122) is often closer to two. It should not be forgotten, however, that the value of $\tilde{C}_{V}$ for polyatomic molecules depends on the internal degrees of freedom that are available at a given temperature (see Chapter 4). Corrections for this effect lead to a significant improvement in the agreement with experimental data, as shown below.

5.3 DIFFUSION

Return now to Fig. 12, again with the layers stationary, and with a uniform concentration in each. Assume that there is a weak concentration gradient between the two. Thus, the z component of the gradient is $(\partial n/\partial z)$ and the difference in concentration between the layers is equal to $\lambda_{o}(\partial n/\partial z)$. The rate of change of concentration as a result of the passage of molecules between the two layers is then equal to

$$\Delta\dot{n} = -2\bar{\bar{z}}\lambda_{o}\left(\frac{\partial n}{\partial z}\right), \tag{123}$$

or

$$\Delta\dot{n} = -D\left(\frac{\partial n}{\partial z}\right). \tag{124}$$

Equation (124) constitutes a definition of the coefficient of diffusion, D, which is then given by

$$D = \frac{1}{\pi\sigma^2 n}\sqrt{\frac{RT}{\pi M}},\tag{125}$$

on the basis of this simple model. This result can also be related to that obtained above for viscosity, since from Eqs (117) and (125),

$$\frac{D}{\eta} = \frac{N_0}{Mn} = \frac{1}{\rho},\tag{126}$$

where ρ is the density of the gas.

5.4 CONCLUSIONS

A summary of the results obtained here for the three most important transport properties is presented in Table 3. The conclusions reached in this Chapter are based on the simple kinetic theory of gases, as initially presented in Chapter 1. The molecules were considered to be point masses, all molecular interactions were neglected and all of the molecules were assumed to have the same speed. This model was sufficient to provide a basis for the analysis of the transport properties of gases. Furthermore, certain approximate relations between the various transport coefficients were developed, e.g. Eqs (121) and (126).

In Chapter 3 the molecules were accorded a certain size. They were treated as rigid spheres of diameter σ, as described in the following chapter. A more rigorous kinetic theory based on this model leads to important modifications in the expressions for the transport coefficients. The results can be expressed as

$$\frac{\eta}{\rho D} = \tfrac{5}{6},\tag{127}$$

$$\frac{\eta \tilde{C}_V}{\kappa M} = \tfrac{2}{5}\tag{128}$$

and

$$\frac{\rho D \tilde{C}_V}{\kappa M} = \frac{\rho D}{\eta} = \tfrac{12}{25}.\tag{129}$$

Table 3 Transport properties of gases

Process	Transport of	Symbol	Simple theoretical expression	Units	
				SI	cgs
Viscous flow	Momentum	η	Eq. (117)	$\mathrm{kg\,m^{-1}\,s^{-1}}$	g/cm-s (poise)
Thermal conduction	Kinetic energy	κ	Eq. (120)	$\mathrm{J\,m^{-1}\,s^{-1}\,K^{-1}}$	ergs/cm-s-degree
Diffusion	Mass	D	Eq. (125)	$\mathrm{m^2\,s^{-1}}$	$\mathrm{cm^2/s}$

Table 4 The Schmidt and Prandtl numbers at 0 °C

Gas	Schmidt Number $\eta/\rho D$	Prandtl Number $\eta \tilde{C}_P/\kappa M$
Ne	0.73	0.66
Ar	0.75	0.67
N_2	0.74	0.71
CH_4	0.70	0.74
O_2	0.74	0.72
CO_2	0.71	0.75
H_2	0.73	0.71
Kinetic theory[a]	0.83	0.67

[a] Rigorous hard-sphere model.

The kinetic theory outlined above suggests that certain ratios of molecular constants should not depend on the nature of the substance. For example, the quantity $\eta \tilde{C}_P/\kappa M$ is known as the Prandlt number. Since the ratio of specific heats is $\gamma = C_P/C_V = 5/3$ for monatomic molecules, Eq. (128) predicts that the Prandlt number should be equal to $2/3$. The experimental values of this quantity for some simple gases shown in Table 4 are in reasonable agreement with this result.

A second parameter that is often used in chemical engineering calculations is the Schmidt number. It is defined by $\eta/\rho D$. From Eq. (127) it would be equal to $5/6$ if the molecules were correctly described by rigid spheres. Some experimental values of this quantity are also given in Table 4. In general, the results of elementary kinetic theory are quite satisfactory for monatomic gases.

In the elementary theory presented above it was assumed that the molecules were monatomic. Thus the rigid-sphere model did not take into account the internal degrees of freedom of diatomic or polyatomic molecules. This question was summarized in Chapter 4. The necessary corrections for the relations between the various transport coefficients were introduced by Euken in the form

$$\kappa = \eta \frac{R}{M} \left(\frac{\tilde{C}_V}{R} + \frac{9}{4} \right), \tag{130}$$

where $\tilde{C}_V$ is a function of temperature, as developed above. The result provides good agreement with the experimental results, although a more sophisticated analysis would require the introduction of intermolecular forces.

In the following chapter the origins of intermolecular forces will be described. Chapter 7 summarizes the properties of real gases as evidence of the practical importance of these forces. Finally, in Chapter 8, these interactions are included in a description of molecular collisions.

Chapter 6

Intermolecular Forces*

The simplest example of an intermolecular interaction is that between two spherical molecules. The force is then a function only of the distance r between their centers. More generally, in a conservative system, the force can be expressed in terms of a scalar interaction potential $\mathscr{V}$. Thus,

$$\mathbf{f} = -\nabla\mathscr{V} = -\operatorname{grad}\mathscr{V} \tag{131}$$

and, for the force between two spherical molecules, Eq. (131) becomes simply

$$f(r) = -\frac{\mathrm{d}\mathscr{V}(r)}{\mathrm{d}r}. \tag{132}$$

The potential function is then given by

$$\mathscr{V}(r) = \int_{r}^{\infty} f(r)\,\mathrm{d}r. \tag{133}$$

A typical potential curve for the interaction of two spherical molecules is shown in Fig. 13. As indicated, it is convenient to separate the intermolecular forces into two types; namely, short-range forces and long-range forces, where the two regions are roughly separated at the minimum value of r.

6.1 ORIGIN OF MOLECULAR INTERACTIONS

6.1.1 Short-range Forces

These forces are often called valence or chemical forces. They arise when the distance r between molecules is short enough so that their electron clouds overlap. These forces are repulsive in nature and in the case of non-spherical molecules may be highly directional.

For spherical molecules, the short-range forces are sometimes represented by

---

*This chapter is based on the presentation by Hirschfelder *et al*. See 'Further Reading' at the end of the book.

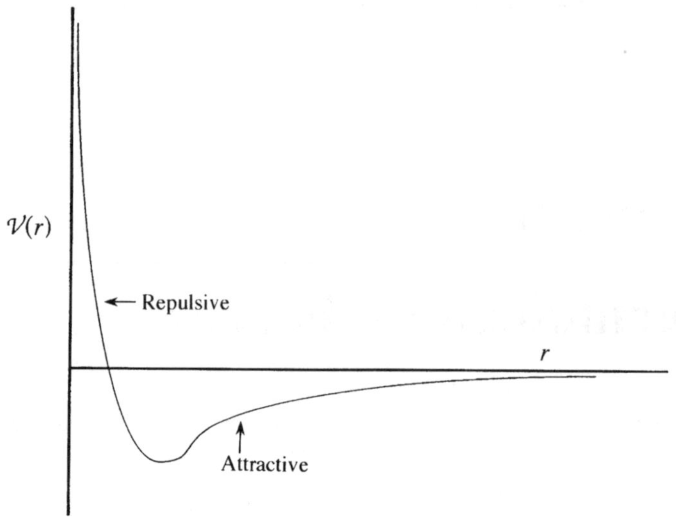

Fig. 13   The intermolecular potential function

an exponential function of the distance of separation, although an inverse power of the distance is more often used for mathematical convenience. Some specific functions will be introduced later in this chapter as descriptions of the short-range forces.

6.1.2   Long-range Forces

These forces vary inversely with the various powers of the intermolecular separation. They can be divided into three types of contribution, namely electrostatic, induction and dispersion. Both electrostatic and induction forces can be treated by classical methods. The dispersion forces, however, are quantum-mechanical in nature. These various contributions will now be briefly summarized.

6.1.2.1   ELECTROSTATIC FORCES

The electrostatic contributions to the intermolecular potential result from the interactions of the various multipole moments in the molecules. These quantities are the charges, dipole moments, quadrupole moments, etc. The presence of dipole, and higher, moments in a molecule precludes its spherical symmetry. Thus, in this case, the more general form of the interaction potential, as given by Eq. (131), must be conserved.

The analytical expressions for the various electrostatic interactions are, in general, very complicated functions of the relative orientations of the two non-spherical molecules. As an important example, it can be shown that the interaction of two point dipoles can be written in the form

$$\mathscr{V}_{ab}^{\cdot(\mu,\mu)} = \frac{\mu_a\mu_b}{r^3}[2\cos\theta_a\cos\theta_b - \sin\theta_a\sin\theta_b\cos(\phi_a - \phi_b)], \qquad (134)$$

where the angles are defined in Fig. 14. However, for many purposes it is sufficient to average the expressions over the angles of orientation. Some results are given in Table 5, where it is apparent that, aside from the case of the interaction of two spherical ions carrying the same charge sign, the averaged potential functions are negative—corresponding to attractive forces. The calculation of these averages involves the Boltzmann factor, which must be included in

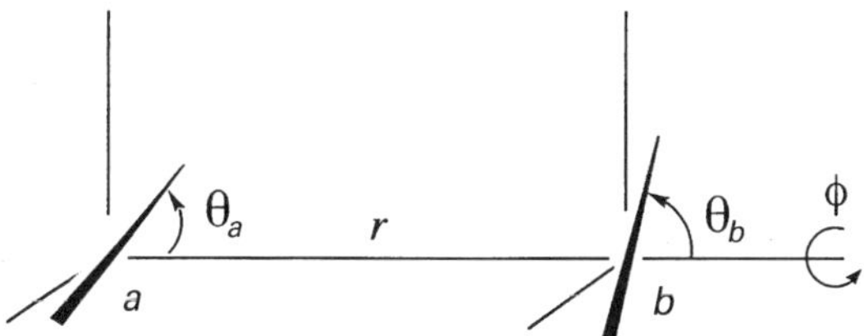

Fig. 14 Coordinates used to specify the interaction between two polar molecules. Note that the angle $\phi = \phi_a - \phi_b$ of Eq. (134) is the dihedral angle between the planes defined by the axes of the two molecules

Table 5 Spherically averaged intermolecular potential functions

Interaction	Effective spherical potential
Charge–charge	$+\dfrac{q_a q_b}{r}$
Charge–dipole	$-\dfrac{1}{3kT}\dfrac{q_a^2\mu_b^2}{r^4}$
Charge–quadrupole	$-\dfrac{1}{20kT}\dfrac{q_a^2 Q_b^2}{r^6}$
Dipole–dipole	$-\dfrac{2}{3kT}\dfrac{\mu_a^2\mu_b^2}{r^6}$
Dipole–induced-dipole	$-\dfrac{\mu_a^2\alpha_b}{2r^6}$
Dipole–quadrupole	$-\dfrac{1}{kT}\dfrac{\mu_a^2 Q_b^2}{r^8}$
Quadrupole–quadrupole	$-\dfrac{7}{40kT}\dfrac{Q_a^2 Q_b^2}{r^{10}}$

order to take into account the statistical distribution of relative molecular orientations. These effective spherical potentials are the result of the assumption that the intermolecular distance does not change significantly during the period of rotation of a given molecule. For relatively large molecular separations, the exponential Boltzmann factor can be expanded; thus, the resulting expressions are temperature-dependent, as shown in Table 5. It should be noted that the first term in the interaction between two neutral molecules is the dipole–dipole term, which varies as r^{-6}.

6.1.2.2 INDUCTION FORCES

The simplest example of electrostatic induction is the interaction of a charged particle, e.g. an ion, with a neutral molecule. If the ion (*a*) of charge q_a and the molecule (*b*) are separated by a distance r, the dipole moment induced in the molecule is given by

$$\mu_b^{(ind)} = \frac{q_a \alpha_b}{r^2},$$ (135)

where α_b is the polarizability of the molecule. This example is illustrated in Fig. 15. The energy of interaction is then equal to

$$\mathscr{V}_{ab}^{(q,\mu^{(ind)})} = \frac{q_a^2 \alpha_b}{r^4}.$$ (136)

The first contribution to the induction energy between neutral particles is due to the interaction of a point dipole in one molecule, say *a*, with the dipole moment that it induces in the other. This dipole – induced-dipole term is of the form

$$\mathscr{V}_{ab}^{(\mu,\mu^{(ind)})} = -\frac{\mu_a^2 \alpha_b (3 \cos^2 \theta_a + 1)}{2 r^6},$$ (137)

where θ_a is the angle between the dipole μ_a and the line of centers between the two molecules. Here, again, it is often useful to average the result over the angle θ_a. The resulting important contribution to intermolecular interactions varies as r^{-6}. It is given by

$$\bar{\mathscr{V}}_{ab}^{(\mu,\mu^{(ind)})} = -\frac{\mu_a^2 \alpha_b}{2 r^6}$$ (138)

(see Table 5).

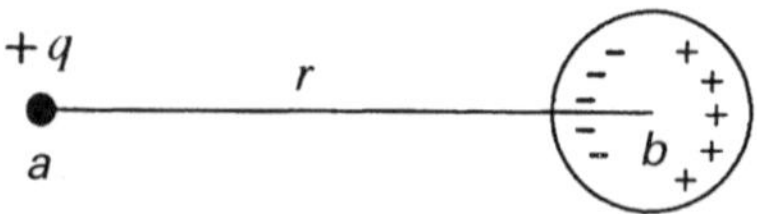

Fig. 15   Induction by a point charge

6.1.2.3 DISPERSION FORCES

Additional long-range, attractive forces are present, even between two non-polar molecules. The quantum-mechanical theory of this type of interaction was developed by London, who showed that the first term is of the form

$$\mathscr{V}_{ab}^{(\text{dis})} = -C\frac{\alpha_a\alpha_b}{r^6}, \tag{139}$$

where constant C, which has dimensions of energy, can be estimated from the ionization potentials of the two molecules, a and b. The dispersion energy given by Eq. (139) is the most important contribution, although in some applications higher terms must be considered.

It will be seen below that various empirical models of intermolecular potential functions have been proposed. The most successful include a term in r^{-6} to represent dipole–dipole, dipole–induced-dipole, and dispersion forces.

6.2 EMPIRICAL POTENTIAL FUNCTIONS

In Chapter 3, the molecules in a gas were accorded a non-zero dimension in order to develop the notion of the mean free path. The collisions of such molecules were then assumed to be elastic, as is the interaction of two billiard balls. The molecules in this case are 'hard spheres' of diameter σ and the corresponding potential function is that shown in Fig. 16(a). The molecular diameter is then a parameter that can be adjusted in numerical calculations to obtain agreement with various experimental results.

The hard spheres can be softened with the use of a potential function that represents the interaction between point-centers of repulsion, as shown in Fig. 16(b). In this case $\mathscr{V}(r) = ar^{-b}$ and two parameters, the force constants a and b, can be varied. A number of properties of gases such as compressibility, viscosity, thermal conductivity, etc. have been treated with the use of this model in order to obtain consistent values of the force constants. It should be noted that no attempt is made with this model to represent attractive forces between molecules.

A combination of the hard-sphere model and a crude attempt to represent the attractive forces is shown in Fig. 16(c). This so-called square-well potential is a function of three adjustable parameters, σ, ε and ϖ. It can thus provide a better fit to experimental data, even though its form does not well represent the molecular interaction. The Sutherland model, as shown in Fig. 16(d), is a combination of hard-sphere repulsion and point-centers of attraction.

The two most realistic potential functions are those proposed by Lennard–Jones and by Buckingham. They are represented in Figs 16(e) and 16(f), respectively. The most general form of the Lennard–Jones potential is

$$\mathscr{V}(r) = \frac{d}{r^\delta} - \frac{c}{r^\gamma}, \tag{140}$$

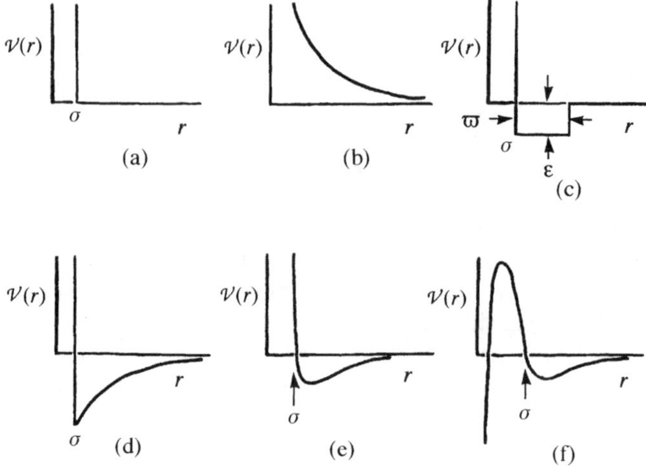

Fig. 16 Various models of spherical molecular interactions: (a) Rigid spheres; (b) Point–centers of repulsion; (c) Square-well potential; (d) Sutherland model; (e) Lennard–Jones potential; (f) Buckingham potential (adapted from Fig. 1.3–3 in *Molecular Theory of Gases and Liquids* by J. G. Hirschfelder, C. F. Curtiss and R. B. Bird, by permission of John Wiley & Sons, Inc., New York, 1954)

where c, d, δ and γ are adjustable parameters. The term d/r^δ represents the repulsive potential, while the term $-(c/r^\gamma)$ corresponds to the attractive potential between the two molecules. The usual form of this function, which is known as the Lennard–Jones (6–12) potential is given by

$$\mathscr{V}(r) = 4\varepsilon\left[\left(\frac{\sigma}{r}\right)^{12} - \left(\frac{\sigma}{r}\right)^{6}\right]. \tag{141}$$

The parameter σ, which is analogous to the molecular diameter introduced earlier, is the value of r for which $\mathscr{V}(r) = 0$. The minimum in the potential-energy curve occurs at $r = \sqrt[6]{\sigma}$, where its depth is equal to ε. It is evident that the second term in Eq. (141) is an attempt to represent the various attractive forces discussed above. The choice of an inverse power term to account for the repulsive forces is more difficult to justify. The exponent 12 is often employed for mathematical convenience, although other inverse powers of r have some-times been used. In summary, the Lennard–Jones (6–12) potential function provides a fairly realistic and reasonably simple representation of the forces between spherical—and hence non-polar—molecules.* Many properties of gases

---

*Molecules are often assumed to preserve their spherical shape, while their polarity is represented by point dipoles. This approximation becomes valid at intermolecular separations that are large com-pared with the molecular dimensions.

have been calculated on the basis of this model and the force constants that characterize this function have been tabulated for many molecular species.

The Lennard–Jones 6–12 potential function is often applied to the collision of two unlike molecules. Thus, in gas mixtures, the heterogeneous interaction *a–b* is described by the force constants

$$\sigma_{ab} = \tfrac{1}{2}(\sigma_a + \sigma_b) \tag{142}$$

and

$$\varepsilon_{ab} = \sqrt{\varepsilon_a \varepsilon_b}, \tag{143}$$

where the subscripts *a* and *b* refer to the force constants for the homogeneous interactions *a–a* and *b–b*, respectively. Equations (142) and (143), which are known as the combining laws for gas mixtures, usually provide satisfactory values for the appropriate force constants.

The second of the more familiar potential functions used to represent the interaction between two spherical molecules is that of Buckingham. In its simplest form it is given by

$$\mathscr{V}(r) = a\,e^{-br} - \left(\frac{\sigma}{r}\right)^6, \tag{144}$$

where it is apparent that the second term corresponds to that employed in the Lennard–Jones (6–12) potential to represent the attractive forces. The exponential first term is suggested from a quantum-mechanical treatment of the short-range forces. Although it has a theoretical basis, it suffers from the fact that at very short distances it turns toward $-\infty$. For many applications, however, this behavior does not pose a serious problem, as the maximum in the curve shown in Fig. 16(f) is often much higher than the relative translational energy of the interacting molecules. In some cases, additional terms in the higher inverse powers of *r* are added to the simple Buckingham (exp-6) function [Eq. (144)].

Finally, it should be mentioned that many attempts have been made to account specifically for the non-spherical form of two interacting molecules. One example is that proposed by Stockmayer, in which the angular dependence of the dipole–dipole interaction, as given in Eq. (134), is added to the Lennard–Jones (6–12) potential of Eq. (141). A number of molecular properties have been evaluated with the use of this function, although it should be obvious that the numerical calculations involved become quite complicated.

6.3 DETERMINATION OF INTERMOLECULAR FORCES

A quantitative evaluation of the force constants for molecular interactions can, in principle, be made from a variety of properties, namely

(i) the equilibrium properties of a gas

(ii)   transport phenomena
(iii) solid-state measurements
(iv)   dynamics studies

The most important equilibrium property of a gas is certainly the relationship between pressure, volume and temperature. The measurement of these quantities yields equation-of-state data that can be compared with various proposed formulas, as presented in Chapter 7. Mention should also be made of the Joule–Thomson effect, for which the measured coefficient can be used to evaluate the intermolecular force constants.

The transport properties of a gas, viscosity, thermal conductivity and diffusion, which were defined in Chapter 5, also provide quantitative information about intermolecular forces. However, because of the numerous approximations, slightly different parameters are often found as compared with those obtained from the analyses of equilibrium properties. In general, the transport phenomena are more sensitive to the repulsive portion of the potential function, while the long-range, attractive part plays a more important role in the equilibrium properties.

The measurement of certain properties of crystals also yields useful data. In particular, the distance between non-polar molecules as the sample temperature approaches absolute zero determines the position of the minimum in the potential curve. Furthermore, the energy of sublimation of the crystal in the low temperature limit yields a value for the depth of the potential well.

Various experiments, which are referred to in (iv) above, include molecular beam measurements, the propagation of sound, the broadening of spectral lines, etc. Some of these dynamical methods will be described in Part II of this book.

Chapter 7

Real Gases

7.1 EQUATIONS OF STATE

The familiar form of the equation of state for an ideal gas is that of Eq. (16). In practical applications to real gases this relation is often generalized with the introduction of the compressibility z. Then, Eq. (16) takes the form

$$PV = z n_{\mathrm{o}} RT, \tag{145}$$

or, in terms of molar volume,

$$z = \frac{P\tilde{V}}{RT}. \tag{146}$$

The compressibility factor has been obtained experimentally for a large number of gases. A few examples of the results are shown in Fig. 17. However, a more

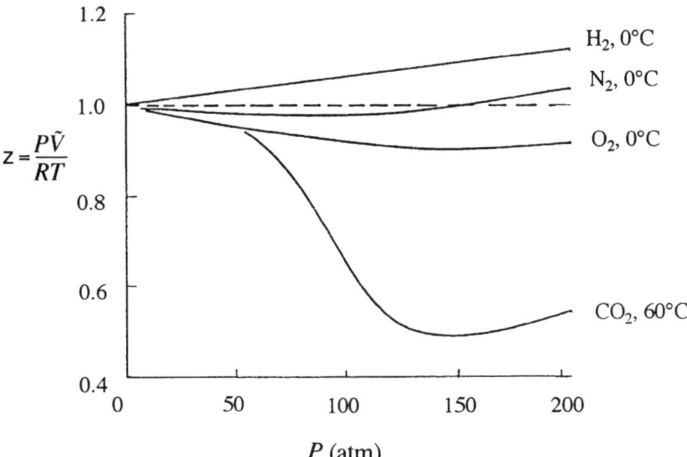

Fig. 17   The compressibility factor as a function of pressure for several gases

fundamental approach to the evaluation of the non-ideal behavior of a gas can be made on the basis of thermodynamic arguments.

The energy per mole of a gas is given by Eq. (44),

$$\tilde{E} = RT^2 \frac{\partial \ln \mathscr{Z}}{\partial T}, \tag{44}$$

where $\mathscr{Z}$ is the partition function. The entropy is defined by

$$S = k \ln \mathscr{W}, \tag{147}$$

where for the non-localized particles in a gas, $\mathscr{W}$ is given by Eq. (29). The definition of the entropy [Eq. (147)] as a logarithmic function of the probability is the result of the fact that it is an additive property of the system. Probabilities, however, combine as the product, as suggested by Eq. (27). It should be noted that the certainty, $\mathscr{W} = 1$, corresponds to $S = 0$. This relation is a statement of the third law of thermodynamics. The fact that k can be identified as the Boltzmann constant is derived from thermodynamic considerations, as shown below.

From Eq. (29)

$$\ln \mathscr{W} = -\sum_i \ln N_i! \tag{148}$$

and, with the aid of Eq. (148), the definition of the partition function and a bit of algebra, the entropy per mole becomes equal to*

$$\tilde{S} = R\left[\ln\left(\frac{\mathscr{Z}}{N}\right) + 1\right] + \frac{\tilde{E}}{T}. \tag{149}$$

The Helmholtz free energy is defined as $\tilde{A} = \tilde{E} - T\tilde{S}$, which with Eqs (44) and (149) leads to the relation

$$\tilde{A} = -RT\left[\ln\left(\frac{\mathscr{Z}}{N}\right) + 1\right]. \tag{150}$$

Finally, the pressure in the system can then be expressed in the form

$$P = -\left(\frac{\partial A}{\partial V}\right)_T = RT\left(\frac{\partial \ln \mathscr{Z}}{\partial V}\right)_T. \tag{151}$$

The resulting thermodynamic or virial equation of state can be written

$$\frac{P\tilde{V}}{RT} = \tilde{V}\left(\frac{\partial \ln \mathscr{Z}}{\partial V}\right)_T, \tag{152}$$

$$= 1 + \frac{B(T)}{\tilde{V}} + \frac{C(T)}{\tilde{V}^2} + \ldots = \mathsf{z}, \tag{153}$$

---

*Note that if the probability distribution given by Eq. (29) is modified to include the degeneracies of the various energy levels, then $\mathscr{W} = \prod_i (g_i^{N_i}/N_i!)$ and Eq. (148) becomes $\ln \mathscr{W} = \sum_i N_i \ln g_i - \sum_i \ln N_i!$. However, the expressions for the various thermodynamic quantities are not modified.

where $B(T)$, $C(T)$, ... are referred to as the second, third, etc. virial coefficients. As an example, for nitrogen at $0\,°C$ the numerical contributions of the virial coefficients to the compressibility factor are found to be

$$1\,\text{atm:}\quad \frac{P\tilde{V}}{RT} = 1 - 0.0005 + 0.000\,003 + \ldots$$

$$10\,\text{atm:}\quad \frac{P\tilde{V}}{RT} = 1 - 0.005 + 0.0003 + \ldots$$

$$100\,\text{atm:}\quad \frac{P\tilde{V}}{RT} = 1 - 0.05 + 0.03 + \ldots$$

The experimental determination of the second and third virial coefficients can be made by measurements of the quantity F, which from Eq. (153) is given by

$$F = \left(\frac{P\tilde{V}}{RT} - 1\right)\tilde{V} = B(T) + \frac{C(T)}{\tilde{V}} + \ldots \tag{154}$$

Thus, if F is plotted as a function of $1/\tilde{V}$, then the intercept is equal to $B(T)$ and the limiting slope determines the value of $C(T)$. Deviations from linearity at lower volumes result from higher terms in the virial expansion (see Fig. 18). It should be noted, however, that the virial expansion is applicable to gases at low and moderate densities, as the series [Eq. (153)] diverges at densities close to that of the liquid.

From Eq. (154) it is evident that departures from real-gas behavior are associated with non-zero values of some or all of the virial coefficients. As a first step in the development of a model of real gases, the molecular diameter was introduced (see Fig. 7(a)). This model corresponds to the potential function of Fig. 16(a) for the interaction of rigid spheres. It is seen from Fig. 7(b) that the

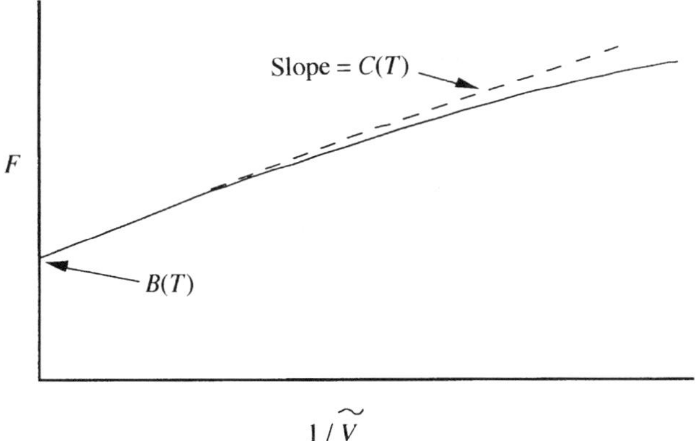

Fig. 18   Determination of the second and third virial coefficients (see text)

effective volume occupied by a pair of identical molecules of diameter σ is equal to $\frac{4}{3}\pi\sigma^3$, or $\frac{2}{3}\pi\sigma^3$ per molecule. The space for molecular movement is then $\tilde{V} - b$ per mole, where $b = N_0\frac{2}{3}\pi\sigma^3$, and the equation of state becomes

$$P(\tilde{V} - b) = RT. \tag{155}$$

The development of Eq. (155) in a series yields

$$\frac{P\tilde{V}}{RT} = \frac{1}{1 - (b/\tilde{V})} = 1 + \frac{b}{\tilde{V}} + \frac{b^2}{\tilde{V}^2} + \ldots, \tag{156}$$

which can be compared with Eq. (153) to give the virial coefficients in terms of the molecular diameter. Thus, $B(T) = b$, $C(T) = b^2$, etc.

A more flexible, and therefore more successful, equation of state is the well-known equation of van der Waals. It is usually written in the form

$$\left(P + \frac{a}{\tilde{V}^2}\right)(\tilde{V} - b) = RT. \tag{157}$$

Its expansion in a manner similar to that of Eq. (156) yields

$$\frac{P\tilde{V}}{RT} = 1 + \frac{b - (a/RT)}{\tilde{V}} + \frac{b^2}{\tilde{V}^2} + \ldots, \tag{158}$$

which, by comparison with Eq. (153), gives the values of the virial coefficients as functions of the parameters a and b in van der Waals' equation, namely $B(T) = b - a/kT$ and $C(T) = b^2$.

To relate the virial coefficients to the parameters employed in a given potential function it is necessary to derive general expressions in terms of the so-called cluster integrals. The reader is referred to more advanced treatments of this problem, as given, for example, by Hirschfelder et al. (see 'Further Reading' at the end of the book). However, in the simple case of angularly independent potential functions, the second virial coefficient can be expressed in the general form

$$B(T) = -\left(\frac{2\pi}{3kT}\right)\int_0^\infty r^3\left(\frac{d\mathscr{V}(r)}{dr}\right)e^{-\mathscr{V}(r)/kT}\,dr. \tag{159}$$

Integration by parts can be employed to transform Eq. (159) into the expression

$$B(T) = 2\pi\int_0^\infty [e^{-\mathscr{V}(r)/kT} - 1]r^2\,dr. \tag{160}$$

Analogous, but much more complicated expressions, can be derived for the third virial coefficient.

The Lennard–Jones (6–12) potential has received particular attention, as it is perhaps the most useful representation of the interaction between spherical molecules. Furthermore, Eq. (160) can, in this case, be evaluated analytically. The result, which is expressed in terms of gamma functions, is given in advanced works on intermolecular forces. This potential function has also been used to calculate the higher virial coefficients in the form of numerical tables.

It is of interest to carry out a simple development of Eq. (160). If the integral is split into two parts, certain approximations can be made. Then, the second virial coefficient for the interaction of spherical molecules can be written in the form

$$B(T) = 2\pi N_{\rm o}\int_0^\sigma (1 - e^{-\mathscr{V}(r)/kT})r^2\,{\rm d}r + 2\pi N_{\rm o}\int_\sigma^\infty (1 - e^{-\mathscr{V}(r)/kT})r^2\,{\rm d}r. \quad (161)$$

Over the region covered by the first integral in Eq. (161), $\mathscr{V}(r) \gg kT$ and the exponential function may be neglected. In the second integral, the expansion of the exponential function yields $1 - e^{-\mathscr{V}(r)/kT} \approx \mathscr{V}(r)/kT$. Thus, Eq. (161) can be approximated by

$$B(T) \approx \left.\frac{2\pi N_{\rm o}r^2}{3}\right]_0^\sigma + \frac{2\pi N_{\rm o}}{kT}\int_\sigma^\infty \mathscr{V}(r)r^2\,{\rm d}r = b - \frac{a}{kT}, \quad (162)$$

where

$$a = -2\pi N_{\rm o}\int_\sigma^\infty \mathscr{V}(r)r^2\,{\rm d}r \quad (163)$$

and

$$b = \frac{2\pi N_{\rm o}\sigma^3}{3}. \quad (164)$$

From this derivation it can be concluded that the parameter b in van der Waals' equation represents the effective molar volume, while the parameter a corresponds to the averaged attractive forces per mole between pairs of interacting molecules.

7.2 CRITICAL PHENOMENA

The equation of van der Waals was introduced above to provide a direct relation between experimental measurements and intermolecular forces. While it is not an accurate representation of equation-of-state data, its beauty lies in its simplicity and the obvious physical significance of the parameters a and b. Furthermore, it can be exploited in the analysis of critical phenomena, as outlined below.

Experimental results of equation-of-state measurements on a given gas are usually expressed as isotherms. Thus, the pressure P, as a function of the molar volume $\tilde{V}$, is determined at a series of fixed temperatures T. Typical results are represented by the curves shown in Fig. 19. At very high temperatures, P and $\tilde{V}$ exhibit the hyperbolic relationship that corresponds to an ideal gas. As the temperature is decreased, the effects of molecular volume and intermolecular forces become apparent, until a temperature T_c is reached at which the curve has a horizontal inflection point. This point is known as the critical point, with pressure P_c and corresponding molar volume $\tilde{V}_c$. The values of the critical constants for several substances are presented in Table 6.

The equation of van der Waals can be differentiated to yield the slope

$$\left(\frac{\partial P}{\partial \tilde{V}}\right)_T = \frac{-RT}{(\tilde{V} - b)^2} + \frac{2a}{\tilde{V}^3}. \tag{165}$$

Note that the slope is equal to zero at infinite molar volume and that it becomes infinite at $\tilde{V} = b$. Furthermore, at the critical point – the point of inflection of

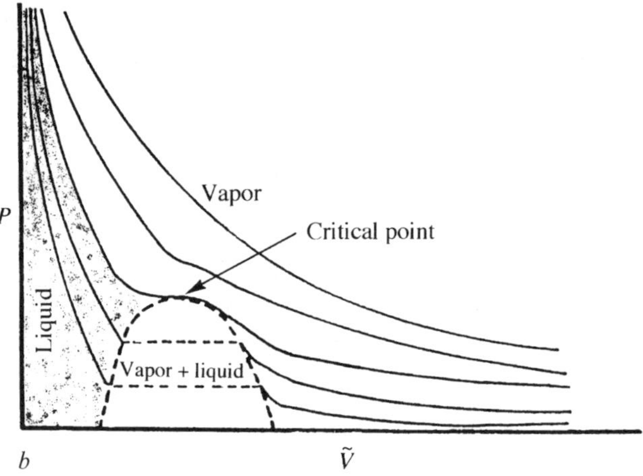

Fig. 19 Typical phase diagram for a neat fluid

Table 6 Critical constants

Substance	T_c (K)	P_c (atm)	$\dfrac{P_c \tilde{V}_c}{RT_c}$
He	5.3	2.26	0.300
Ne	44.5	25.90	0.297
Ar	151.2	48.00	0.292
Kr	210.2	54.00	0.336
Xe	289.8	58.20	0.277
H_2	33.3	12.80	0.304
N_2	126.1	33.50	0.293
O_2	154.4	49.70	0.292
CO	134.2	35.00	0.286
CO_2	304.3	73.00	0.280
NO	179.2	65.00	0.256

the curve T_c, the slope is also equal to zero. This condition becomes

$$\frac{2a}{RT_c} = \frac{\tilde{V}_c^3}{(\tilde{V}_c - b)^2}. \tag{166}$$

The derivative of Eq. (165), which is given by

$$\left(\frac{\partial^2 P}{\partial \tilde{V}^2}\right)_T = \frac{2RT}{(\tilde{V} - b)^3} - \frac{6a}{\tilde{V}^4}, \tag{167}$$

is also equal to zero at the inflection point. Thus,

$$\frac{3a}{RT_c} = \frac{\tilde{V}_c^4}{(\tilde{V}_c - b)^3}, \tag{168}$$

and division of Eq. (166) by Eq. (168) leads to the relations

$$\frac{2}{3} = \frac{\tilde{V}_c - b}{\tilde{V}_c} \tag{169}$$

and

$$\tilde{V}_c = 3b. \tag{170}$$

Substitution of this result in van der Waals' equation yields

$$T_c = \frac{8a}{27bR} \tag{171}$$

or

$$P_c = \frac{a}{27b^2}. \tag{172}$$

By combining Eqs (168)–(170) it is easily found that

$$\frac{P_c \tilde{V}_c}{RT_c} = \frac{3}{8} = 0.375. \tag{173}$$

This value should, of course, be the same for any gas that obeys van der Waals' equation. However, from Table 6 it is evident that the experimental values of this quantity are somewhat lower. In particular, for substances in which the inter-molecular interactions are strong because of hydrogen-bond formation (e.g. water, methyl alcohol) the value is significantly lower. Furthermore, it should be pointed out that other equations of state can yield better values than that given in Eq. (173). For example, Dieterichi's equation* leads to 0.271, which is in better agreement with the values given in Table 6.

---

*Dieterichi's equation can be written in the form

$$P = \left(\frac{n_0 RT}{V - bn_0}\right) e^{-an_0/VRT},$$

where a and b are constants;

For convenience, van der Waals' equation will now be used to illustrate a general principle, that of corresponding states. The parameters a and b can be evaluated from Eqs (170) and (172), which yield

$$b = \tfrac{1}{3}\tilde{V}_{\rm c} \qquad (174)$$

and

$$a = 3P_{\rm c}\tilde{V}_{\rm c}^2. \qquad (175)$$

Substitution into van der Waals' equation [Eq. (157)] leads to its reduced form, namely

$$\left(P_{\rm R} + \frac{3}{\tilde{V}_{\rm R}^2}\right)(3\tilde{V}_{\rm R} - 1) = 8T_{\rm R}, \qquad (176)$$

in terms of the reduced variables $P_{\rm R} = P/P_{\rm c}$, $\tilde{V}_{\rm R} = \tilde{V}/\tilde{V}_{\rm c}$ and $T_{\rm R} = T/T_{\rm c}$. It should be noted that Eq. (176) contains no arbitrary parameters. Analogous reduced equations can be derived for all gases that exhibit a critical point and obey a given two-parameter equation of state. Clearly, more than two parameters cannot be evaluated from the experimental determination of the critical point.

7.3 TRANSPORT PROPERTIES

The transport properties of ideal gases were presented in Chapter 5. The treatment was based on the model of the ideal gas presented in Chapter 1. However, it was necessary to define the non-zero dimension of the particle in order to introduce the concept of the mean free path. Other contributions to intermolecular interactions were not considered.

With the summary of intermolecular forces in Chapter 6, it is now of interest to reconsider the transport properties of a gas, albeit from a qualitative point of view. As the molecules of a gas move about, they collide with one another, resulting in an exchange of both energy and momentum. These collisions serve as an equilibrating mechanism in that the less energetic partners tend to gain energy at the expense of loss by the other, more energetic, molecules in the system. When the average properties of each of the colliding molecules are the same, no average changes occur and a uniform steady state of the gas results. However, in a so-called non-uniform gas, if the molecules in collision originate from regions with different average properties of energy and momentum, the net effect of the collisions is to provoke a transfer of these quantities. The continuation of such a transfer must be assured by the application of external forces.

As an example of a transport property of a real, dilute gas, its viscosity will now be developed with the use of a simple intermolecular potential function. It was found in Chapter 5 that, on the basis of the ideal-gas formalism, the viscosity of a gas would be expected to be proportional to the square root of the temperature [see Eq. (117)]. However, measurements at constant pressure show that, in fact, the viscosity increases more rapidly with temperature. A dynamical model that is consistent with this observation requires the introduction of intermolecular forces.

Figure 16(d) represents the Sutherland model, a spherically symmetric potential function consisting of hard-sphere repulsion and a simple inverse-power attraction. As shown in the following chapter, on the basis of this model the effective molecular diameter σ' becomes a function of g, the relative kinetic energy of the colliding molecules. Thus, it is found that

$$\sigma'^2 = \sigma^2 \left[1 - \frac{\mathscr{V}(\sigma)}{\frac{1}{2}\mu g^2} \right], \tag{177}$$

where $\frac{1}{2}\mu g^2$, the relative kinetic energy of two colliding molecules, is proportional to T. Combination of Eqs (117) and (177) leads to a semi-empirical expression for the temperature dependence of the viscosity; namely,

$$\eta = \frac{M}{\pi N_0 \sigma^2} \sqrt{\frac{RT}{\pi M}} \left(1 + \frac{s}{T} \right)^{-1}, \tag{178}$$

where s is an adjustable parameter. Equation (178) has been employed with success to represent the temperature dependence of the viscosity of many gases over a wide range of temperature.

Chapter 8

Molecular Collisions

In the elementary kinetic theory of gases presented in Chapter 1, the nature of molecular collisions was not considered. Nevertheless, some important properties of gases were developed—in particular, the ideal-gas law. With the description of intermolecular forces introduced in Chapter 6 it becomes possible to treat the problem of molecular collisions in a more specific way. However, it is still difficult to develop the problem in general. In the following chapters, therefore, the analysis will be limited to the binary interactions of spherical molecules. The results obtained yield an interesting and useful semiquantitative picture of molecular interactions. Two fundamentally different models will be described: namely, that based on classical mechanics and the other, derived from Schrödinger's equation, that requires a certain knowledge of elementary quantum mechanics.

8.1 BINARY COLLISIONS IN CLASSICAL MECHANICS

In the classical treatment of bimolecular collisions, the angle of deflection is the only necessary characteristic of the encounter. Thus, for example, this quantity can be employed to evaluate the various transport phenomena as functions of a given intermolecular potential function.

Consider a system composed of only two molecules, one of species a with mass m_a and the other, species b with mass m_b. The interaction force is assumed to depend only on the distance between them. Initially, before a collision takes place, the molecular velocities are given by $\mathbf{u}_a$ and $\mathbf{u}_b$, respectively. After a collision, the corresponding masses are m'_a and m'_b and the velocities are represented by $\mathbf{u}'_a$ and $\mathbf{u}'_b$. With the application of the laws of conservation of mass, linear momentum and energy, certain relations known as the invariants of the encounter can be specified.

If no chemical reaction occurs between the colliding molecules, then the conservation of mass of the system leads to the relations

$$m_a = m'_a, \qquad m_b = m'_b \tag{179}$$

and thus

$$m_a + m_b = m'_a + m'_b. \tag{180}$$

If no external forces act on the system, then the conservation of linear momentum is expressed by

$$m_a \mathbf{u}_a + m_b \mathbf{u}_b = m_a \mathbf{u}'_a + m_b \mathbf{u}'_b. \tag{181}$$

At any time during the collision, the sum of the kinetic energies of the particles and their energy of interaction is constant. Before and after the encounter the particles are separated by an infinite distance. Hence, the potential energy is equal to zero and the total energy is just the sum of the kinetic energies of the particles. This condition for the conservation of energy can be written as

$$\tfrac{1}{2}m_a u_a^2 + \tfrac{1}{2}m_b u_b^2 = \tfrac{1}{2}m'_a u'^2_a + \tfrac{1}{2}m'_b u'^2_b. \tag{182}$$

If the particles are spherical, then the three conditions expressed by Eqs (180)–(182) are sufficient to determine precisely the trajectory of the collision. The result is given by a formula for the angle of deflection, a quantity that enters directly in the equations that describe the transport phenomena. In the development, the problem is first reduced to that of the motion of two bodies in a plan and, then, to the motion of a hypothetical particle in two dimensions.

Newton's second law of motion can be applied in the form

$$\mathbf{f}_a = m_a \frac{\mathrm{d}^2 \mathbf{r}_a}{\mathrm{d}t^2} \tag{183}$$

and

$$\mathbf{f}_b = m_b \frac{\mathrm{d}^2 \mathbf{r}_b}{\mathrm{d}t^2} = -\mathbf{f}_a, \tag{184}$$

where $\mathbf{f}_a$ and $\mathbf{f}_b$ are the forces acting on particles a and b, respectively, and $\mathbf{r}_a$ and $\mathbf{r}_b$ are the corresponding position vectors. The second equality in Eq. (184) specifies that the only force acting on the molecules is that between them. Equations (183) and (184) can be combined to yield

$$\frac{\mathrm{d}^2}{\mathrm{d}t^2}(\mathbf{r}_a - \mathbf{r}_b) = \frac{\mathbf{f}_a}{\mu}, \tag{185}$$

where μ is the reduced mass of the colliding pair, defined by

$$\frac{1}{\mu} = \frac{1}{m_a} + \frac{1}{m_b} \tag{186}$$

[see Eq. (97)].

The vector product of Eq. (185) with the vector $(\mathbf{r}_a - \mathbf{r}_b)$ is equal to zero, as for these spherical particles the force acts along the line of centers. Thus,

$$(\mathbf{r}_a - \mathbf{r}_b) \times \frac{\mathrm{d}^2}{\mathrm{d}t^2}(\mathbf{r}_a - \mathbf{r}_b) = 0 \tag{187}$$

or

$$\frac{d}{dt}\left[(\mathbf{r}_a - \mathbf{r}_b) \times \frac{d}{dt}(\mathbf{r}_a - \mathbf{r}_b)\right] - \left[\frac{d}{dt}(\mathbf{r}_a - \mathbf{r}_b) \times \frac{d}{dt}(\mathbf{r}_a - \mathbf{r}_b)\right] = 0. \qquad (188)$$

The second term on the left-hand side of Eq. (188) is, of course, equal to zero, while the first term can be integrated with respect to time to yield

$$(\mathbf{r}_a - \mathbf{r}_b) \times \frac{d}{dt}(\mathbf{r}_a - \mathbf{r}_b) = (\mathbf{r}_a - \mathbf{r}_b) \times (\mathbf{u}_a - \mathbf{u}_b) = \mathbf{C}, \qquad (189)$$

where $\mathbf{C}$ is a constant vector. Note that $\mathbf{C}$ is perpendicular to the plane formed by the vectors $(\mathbf{r}_a - \mathbf{r}_b)$ and $(\mathbf{u}_a - \mathbf{u}_b)$. Thus, at all times, the two particles, as well as the center of mass of the system, lie in a plane normal to $\mathbf{C}$

On the basis of the argument presented above, it becomes possible to describe the dynamics of the two-particle system with the use of two pairs of coordinates x_a, y_a and x_b, y_b, where the z axis is chosen in the direction of $\mathbf{C}$ and, for convenience, the origin is placed at the center of mass of the system. This result is illustrated in Fig. 20. The kinetic energy can then be written in the form

$$\mathscr{T} = \tfrac{1}{2}m_a(\dot{x}_a^2 + \dot{y}_a^2) + \tfrac{1}{2}m_b(\dot{x}_b^2 + \dot{y}_b^2). \qquad (190)$$

Equation (190) can be rewritten in polar coordinates. Furthermore, as the origin is the center of mass, the coordinates $x_a y_a$ and $x_b y_b$ can be transformed in terms of r and θ via the substitutions

$$x_a = \frac{m_b}{m_a + m_b} r \cos \theta, \qquad (191)$$

$$y_a = \frac{m_b}{m_a + m_b} r \sin \theta, \qquad (192)$$

$$x_b = \frac{m_a}{m_a + m_b} r \cos \theta \qquad (193)$$

and

$$y_b = \frac{m_a}{m_a + m_b} r \sin \theta. \qquad (194)$$

The kinetic energy expressed in center-of-mass, polar coordinates then takes the form

$$\mathscr{T} = \tfrac{1}{2}\mu(\dot{r}^2 + r^2\dot{\theta}^2). \qquad (195)$$

This relation is identical to that obtained for the kinetic energy of a single particle of mass μ moving under the influence of a spherically symmetric potential field. The total energy is now written as the sum of Eq. (195) and the appropriate potential function $\mathscr{V}(r)$.

It is useful to define two parameters that, with the potential function, characterize the collision, namely

(i)   The impact parameter b, which is the distance of closest approach in the absence of the potential (see Fig. 21), and

(ii) The initial relative speed g of the colliding particles.

Before the advent of the collision ($r = \infty$) the potential is equal to zero and the kinetic energy $\frac{1}{2}\mu g^2$ is the total energy of the system. Furthermore, the angular moment is given by $\mu b g$. Thus, the conservation of energy and angular momentum throughout the collision can be written as

$$\tfrac{1}{2}\mu g^2 = \tfrac{1}{2}\mu(\dot{r}^2 + r^2\dot{\theta}^2) + \mathscr{V}(r) \tag{196}$$

and

$$\mu b g = \mu r^2 \dot{\theta}, \tag{197}$$

where the right-hand side of Eq. (197) is obtained by taking the derivative of Eq. (195) with respect to $\dot{\theta}$.

Equations (196) and (197) can be combined by eliminating $\dot{\theta}$ to yield

$$\tfrac{1}{2}\mu g^2 = \tfrac{1}{2}\mu\dot{r}^2 + \tfrac{1}{2}\mu g^2\left(\frac{b}{r}\right)^2 + \mathscr{V}(r), \tag{198}$$

which allows r to be determined as a function of time, i.e. the trajectory of the collision.

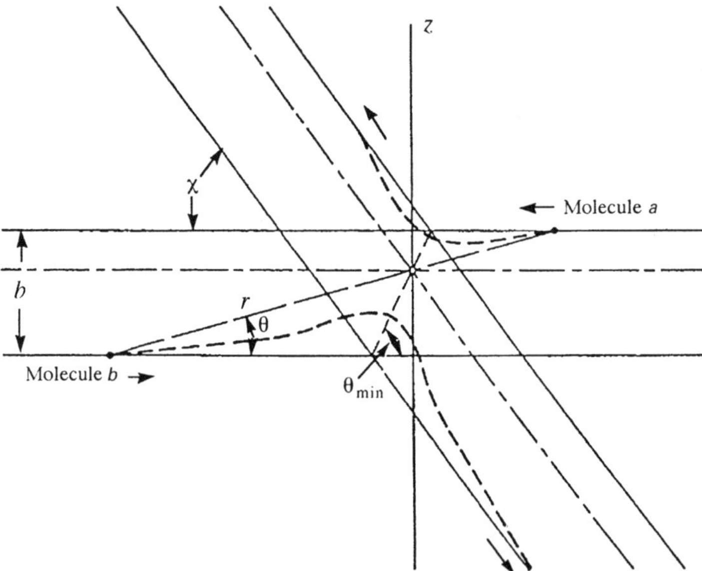

Fig. 20   A binary collision in the plane of interaction (adapted from Fig. 1.5-2 in *Molecular Theory of Gases and Liquids* by J. G. Hirschfelder, C. F. Curtiss and R. B. Bird, by permission of John Wiley & Sons, Inc., New York, 1954)

It should be noted that Eq. (198) is independent of θ; thus, it describes a one-dimensional motion of a particle of mass μ with a total energy $\frac{1}{2}\mu g^2$ in an effective potential given by

$$\mathscr{V}_{\text{eff}}(r) = \mathscr{V}(r) = \tfrac{1}{2}\mu g^2 \left(\frac{b}{r}\right)^2 . \tag{199}$$

The second term on the right-hand side of Eq. (199) is referred to as the centrifugal potential. An example is shown in Fig. 22. It should be emphasized that this figure represents only one in a family of curves, depending on the values of the initial parameters b and g.

The most important element in the description of a binary collision is the angle of deflection, χ. This quantity is defined in Figs 20 and 21; it is there related to the angle $\theta_{\min}$ by

$$\chi = \pi - 2\theta_{\min}. \tag{200}$$

As shown in Fig. 21, $\theta_{\min}$ is the value of the angle θ for which r has a minimum value $r_{\min}$, the distance of closest approach.

The angle $\theta_{\min}$ can be calculated, since from Eqs (197) and (198), r and θ are given as functions of time. Then,

$$\frac{\mathrm{d}r}{\mathrm{d}\theta} = \frac{\dot{r}}{\dot{\theta}} = -\left(\frac{r^2}{b}\right)\sqrt{1 - \frac{\mathscr{V}(r)}{\frac{1}{2}\mu g^2} - \frac{b^2}{r^2}}. \tag{201}$$

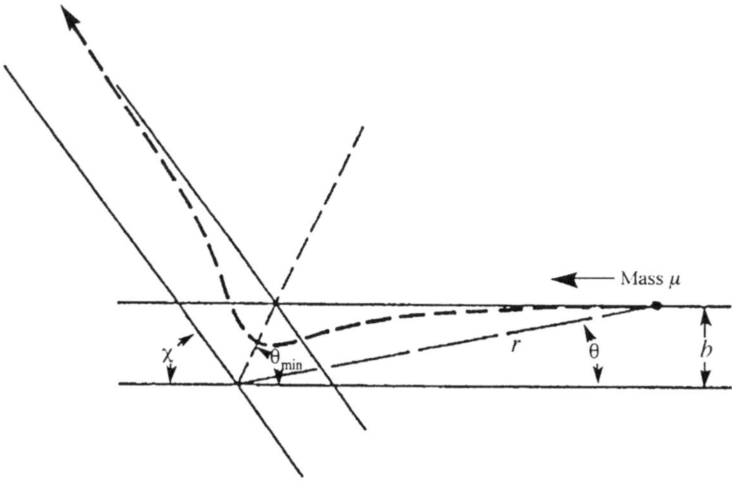

Fig. 21 A binary collision relative to the center of mass of the system (adapted from Fig. 1.5-3 in *Molecular Theory of Gases and Liquids* by J. G. Hirschfelder, C. F. Curtiss and R. B. Bird, by permission of John Wiley & Sons, Inc., New York, 1954)

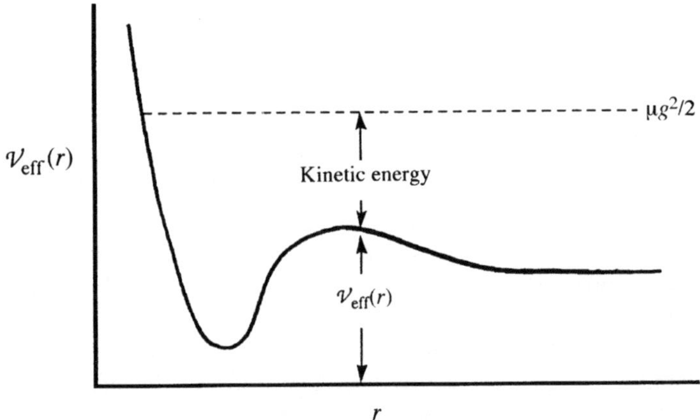

Fig. 22   Effective potential energy for the collision of two spherical molecules

Note that $dr/d\theta < 0$ has been chosen, so that r decreases with θ along the incoming trajectory. The angle $\theta_{\min}$ can then be calculated from the integral

$$\theta_{\min} = \int_0^{\theta_{\min}} d\theta = -b\int_\infty^{r_{\min}} \frac{1}{\sqrt{1 - (b^2/r^2) - [\mathscr{V}(r)/\frac{1}{2}\mu g^2]}}\frac{dr}{r^2}. \tag{202}$$

Then, from Eq. (200), the angle of deflection is given by

$$\chi(b,\,g) = \pi - 2b\int_{r_{\min}}^\infty \frac{1}{\sqrt{1 - (b^2/r^2) - [\mathscr{V}(r)/\frac{1}{2}\mu g^2]}}\frac{dr}{r^2}. \tag{203}$$

This expression for the angle of deflection, which results from a binary collision defined by the values of b and g, is valid for any spherically symmetric potential function $\mathscr{V}(r)$. The lower limit on the integral of Eq. (202) is evaluated by setting the right-hand side of Eq. (201) equal to zero.

For dilute, real gases, where ternary and higher collisions can be neglected, Eq. (203) provides the basic classical relationship, which can be used to evaluate a number of properties. As examples, the second virial coefficient is given by

$$B(T) = \tfrac{4}{3}N_0\sqrt{\pi}\int_0^\infty e^{-\gamma^2}\gamma^4\left[\int_0^\infty \chi\,db^3\right]d\gamma, \tag{204}$$

where $\gamma^2 = \frac{1}{2}\mu g^2/2kT$. The analogous expression,

$$\frac{RT}{\eta(T)} = \tfrac{4}{5}N_0\sqrt{\pi}\int_0^\infty e^{-\gamma^2}\gamma^7\left[\int_0^\infty \sin^2\chi\,db^2\right]d\gamma, \tag{205}$$

can be used to evaluate the coefficient of viscosity from the angle of deflection.

The other transport properties can also be calculated from Eq. (203), although the necessary formulas are even more complicated. The interested reader is referred to advanced books on this subject.

8.2 QUANTUM THEORY OF BINARY COLLISIONS

In the classical picture of two-particle interaction outlined above, it was shown that a specific quantity—the angle of deflection—characterizes a given collision. However, on the atomic-molecular scale, quantum theory is more appropriate. According to the uncertainty principle of Heisenberg, the simultaneous determination of the position and momentum of a particle cannot be made. Thus, it is not possible to determine exactly the angle of deflection in a collision. In the following development it is found that the phase shift of the radial wave function characterizes a binary, quantum-mechanical collision. This quantity, then, which is analogous to the classical angle of deflection, determines the final quantum-mechanical expressions for the second virial coefficient and the low-density transport coefficients for low-pressure real gases.

It is assumed here that the reader has a knowledge of basic quantum mechanics. The treatment of two-particle, bound systems such as the hydrogen atom and the harmonic vibration of a diatomic molecule are invariably presented in elementary courses. However, the solution to the binary collision problem, as outlined below, is based on the WKB method (after Wentzel, Kramers and Brillouin), which is not usually given. It will be presented here in a very simplified form in order to introduce the concept of the phase shift.

It was shown in Chapter 7 that a two-body problem can be reduced to the movement of a hypothetical particle of mass μ in three dimensions. The Schrödinger equation can then be written in the form

$$-\frac{h^2}{8\pi^2\mu}\nabla^2\psi + \mathscr{V}(r)\psi = \varepsilon\psi, \tag{206}$$

where ∇^2 is the Laplacian operator, h is Planck's constant and ψ is the wave function that provides the quantum-mechanical description of the binary collision. With the use of spherical, center-of-mass coordinates, the Laplacian operator is given by

$$\nabla^2 = \frac{1}{r^2}\frac{\partial}{\partial r}\left(r^2\frac{\partial}{\partial r}\right) + \frac{1}{r^2\sin^2\theta}\frac{\partial^2}{\partial\varphi^2} + \frac{1}{r^2\sin\theta}\frac{\partial}{\partial\theta}\left(\sin\theta\frac{\partial}{\partial\theta}\right) \tag{207}$$

and the wave function is $\psi(r, \theta, \varphi)$. The separation of variables is now carried out in the usual way, namely $\psi(r, \theta, \varphi) = R(r)\Theta(\theta)\Phi(\varphi)$. Thus, the angular part of the wave function can be expressed in spherical harmonics in the form

$Y_\ell^m(\theta, \varphi)$.* The quantum numbers ℓ and m, which correspond to those of the hydrogen-atom problem, were discussed in Chapter 4.

The remaining equation in $R(r)$, which is of interest in the present problem,

$$-\frac{h^2}{8\pi^2\mu}\frac{1}{r^2}\frac{\mathrm{d}}{\mathrm{d}r}\left(r^2\frac{\mathrm{d}R}{\mathrm{d}r}\right) + \frac{h^2\ell(\ell+1)}{8\pi^2\mu}R + \mathscr{V}(r)R = \varepsilon R, \tag{208}$$

can be solved for certain forms of the potential function, $\mathscr{V}(r)$. It should be noted that the second term in Eq. (208) corresponds to the centrifugal energy, as given in the classical treatment by the second term in Eq. (199). However, in quantum mechanics this term is a function of the quantum number $\ell = 0, 1, 2, \ldots$; thus, the angular momentum is quantized, unlike in the classical example. In general, the substitution $R(r) = S(r)/r$ in Eq. (208) yields the simpler form of this second-order differential equation, namely

$$\frac{\mathrm{d}^2 S(r)}{\mathrm{d}r^2} + \left\{-\frac{\ell(\ell+1)}{r^2} + \frac{8\pi^2\mu}{h^2}[\varepsilon - \mathscr{V}(r)]\right\}S(r) = 0. \tag{209}$$

Now, consider the case of an ideal gas, for which $\mathscr{V}(r) = 0$. Equation (209) can then be written as

$$\frac{\mathrm{d}^2 S^0(r)}{\mathrm{d}r^2} + \left\{-\frac{\ell(\ell+1)}{r^2} + \frac{4\pi^2\mu^2 g^2}{h^2}\right\}S^0(r) = 0, \tag{210}$$

where $\mathscr{V}(r) = 0$ and the energy is equal to $\mu g^2/2$. This relation can be compared with the general equation of Bessel,

$$\frac{\mathrm{d}^2 S^0(r)}{\mathrm{d}r^2} + \left(\beta^2 - \frac{4p^2 - 1}{4r^2}\right)S^0(r) = 0. \tag{211}$$

Thus, $\beta^2 = 4\pi^2\mu^2 g^2/h^2$ and the index p can be determined from the relation $(4p^2 - 1)/4 = \ell(\ell+1)$, or $p = \pm(\ell + \frac{1}{2})$. The general solution to Bessel's equation is of the form

$$Z_p(\beta r) = A_\ell(\beta)J_{\ell+(1/2)}(\beta r) + B_\ell(\beta)N_{-\ell-(1/2)}(\beta r), \tag{212}$$

where $J_{\ell+(1/2)}(\beta r)$ and $N_{-\ell-(1/2)}(\beta r)$ are the Bessel and von Neumann functions, respectively. If Eq. (212) is not familiar to the reader, then it should be noted that it is analogous to the general solution to the usual second-order differential equation for harmonic motion, which is written as a linear combination of sine and cosine functions. In both cases, the coefficients are determined

---

*The spherical harmonics are defined as a normalized product of the associated Legendre polynomials and an exponential function, namely

$$Y_\ell^m(\theta, \phi) = P_\ell^{|m|}(\cos\theta)\,\mathrm{e}^{\mathrm{i}m\phi},$$

where the polar angles θ and ϕ are defined in Fig. I.4, with $\ell = 0, 1, 2, \ldots$ and $m = 0, \pm 1, \pm 2, \ldots, \pm\ell$.

Table 7 Characteristics of a binary collision

Classical mechanics	Quantum mechanics
$\chi(b, g)$: Angle of deflection	$\eta_\ell(\beta)$: Phase shift
b: Impact parameter	ℓ: Angular momentum quantum number
$\mu g b$: Angular momentum	$\hbar\sqrt{\ell(\ell+1)}$: Angular momentum
g: Initial relative speed	$\beta = \mu g/\hbar = 2\pi/\lambda$: Wavenumber of the deBroglie wave
μg: Relative momentum	$\beta\hbar$: Relative momentum

by the appropriate boundary conditions. In Eq. (212), $B_\ell(\beta)$ must be set equal to zero, since the von Neumann function becomes infinite in the limit as $r \to \infty$. Furthermore, the Bessel function $J_{\ell+(1/2)}(\beta r)$ becomes sinusoidal for large values of r, namely $S^0 = rR^0 \to A_\ell(\beta)\sin(\beta r - \frac{1}{2}\pi\ell)$, a representation of the deBroglie wave of the hypothetical particle of mass μ.

Returning now to the more general problem where there is an interaction potential $\mathscr{V}(r)$, the solutions to Eq. (209) can also be expressed by Eq. (212). However, with $\mathscr{V}(0) \neq 0$, the previous boundary condition is no longer applicable. Thus, the coefficient $B_\ell(\beta) \neq 0$, and for large values of r the asymptotic solution is given by

$$S(r) = rR(r) = [A_\ell^2(\beta) + B_\ell^2(\beta)]^{1/2} \sin[\beta r - \tfrac{1}{2}\pi\ell + \eta_\ell(\beta)]. \tag{213}$$

The argument of sine in Eq. (213) now contains the phase shift,

$$\eta_\ell(\beta) = \tan^{-1}[(-1)^\ell B_\ell(\beta)/A_\ell(\beta)], \tag{214}$$

which results from the molecular interaction. In the quantum-mechanical picture of the encounter it is the phase shift that is characteristic. This quantity is analogous to the angle of deflection in the classical case (see Table 7).

Part II
APPLICATIONS

Chapter 9

Effusion and the Separation of Mixtures

An elementary treatment of the collision of a molecule with a wall was presented in Chapter 1. This simple model was sufficient for the development of the ideal-gas law. The molecules were treated as hard spheres in the discussion of transport properties, although it was pointed out that a more rigorous analysis of molecular collisions had to be made in order to obtain reasonable agreement with experimental results. In the present application, the collision of a molecule with a wall will be described in somewhat more detail.

To calculate the number of molecules striking a wall (say, in the x, y plane) in unit time, it is necessary to take into account not only the distribution of molecular speeds, as derived in Chapter 3, but also the distribution of directions in space (see Fig. 1). For present purposes, it will be assumed that the wall is perfectly smooth on the molecular scale and that an element of the surface dS can be defined. In an elastic collision between a hard-sphere molecule at the left

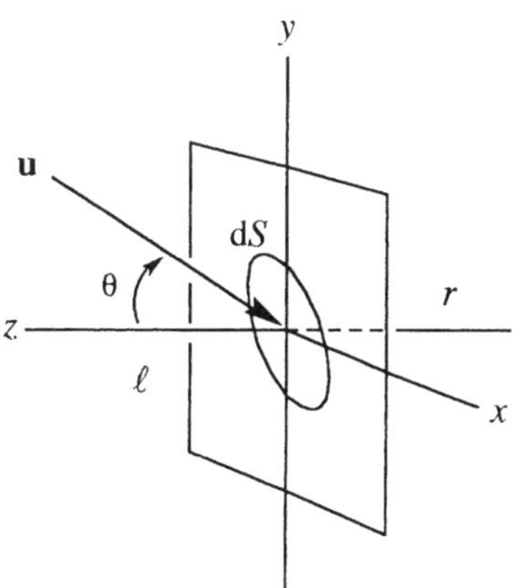

Fig. 1 Collision of a molecule of velocity **u** with a surface in the x, y plane

of the wall and the surface element dS, the forces at contact are normal to the surface. Thus, the component of the velocity parallel to the surface is not altered by the collision. However, as the collisions are assumed to be elastic, the normal component of the velocity is reversed. The molecule of mass m is thus 'specularly reflected', by analogy with Snell's law in optics (see Fig. 2), and the change in momentum of the molecule is along the z direction. Since the normal component of the initial momentum is equal to $-m\dot{z}$ and the final component is $+m\dot{z}$, the momentum transferred to the wall is given by $2m\dot{z}$.

From Eq. (I.51), of a total of N molecules the number with speeds between u and $u + \mathrm{d}u$ is given by

$$\mathrm{d}N = 4\pi N \left(\frac{m}{2\pi kT}\right)^{3/2} \mathrm{e}^{-mu^2/2kT} u^2 \, \mathrm{d}u. \tag{1}$$

The direction of the velocity vector $\mathbf{u}$ is conveniently specified in spherical coordinates, as defined in Fig. I.4. The fraction of molecules whose velocities are within the solid angle $\mathrm{d}\Omega = \sin\theta \, \mathrm{d}\theta \, \mathrm{d}\phi$ is equal to $\mathrm{d}\Omega/4\pi$, as $\Omega = 4\pi$ is the solid angle subtended by the surface of a sphere. Initially, all of these molecules are contained in an oblique cylinder of base dS and length $u \, \mathrm{d}t$, whose axis is parallel to $\mathbf{u}$, as shown in Fig. 3. The interval of time dt is assumed to be short enough so that the distance $u \, \mathrm{d}t$ is small with respect to the mean-free path for successive collisions of the molecule. The total number of molecules hitting the surface element dS from all directions during the time interval dt is then equal to

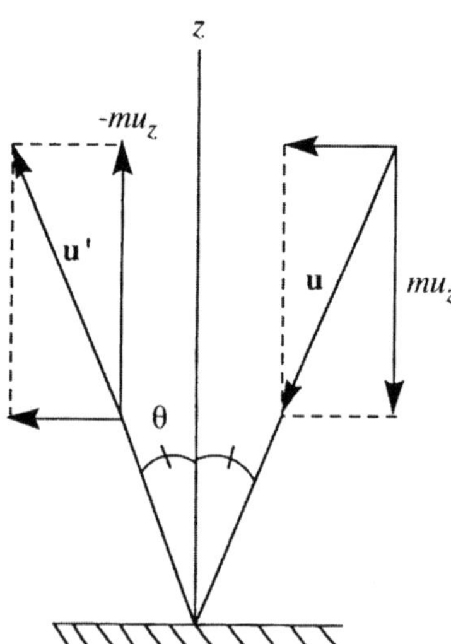

Fig. 2   Change in the z component of the momentum upon collision with the surface

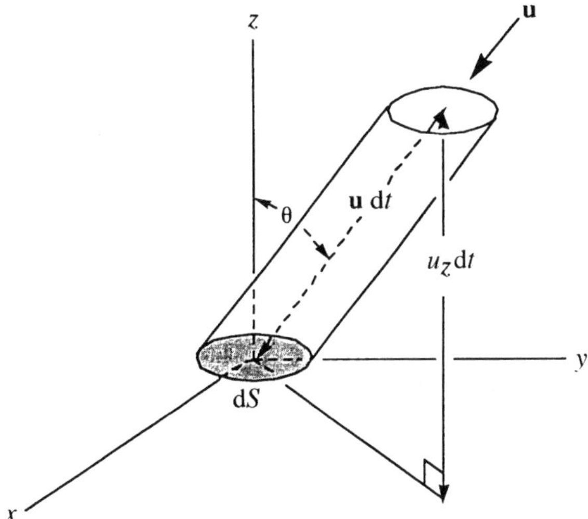

Fig. 3 Oblique cylinder containing molecules of velocity **u** that strike the surface element dS in the time interval dt

$$\left[4\pi N\left(\frac{m}{2\pi kT}\right)^{3/2}e^{-mu^2/2kT}u^2\,\mathrm{d}u\right]\left(\frac{\sin\theta\,\mathrm{d}\theta\,\mathrm{d}\phi}{4\pi}\right)(u\cos\theta\,\mathrm{d}S\,\mathrm{d}t).$$

The integration over the possible directions of **u** is easily carried out; namely,

$$\int_0^{\pi/2}\int_0^{2\pi}\cos\theta\sin\theta\,\mathrm{d}\theta\,\mathrm{d}\phi = \pi. \tag{2}$$

However, as molecules strike only the left side of the wall (ℓ in Fig. 1), the limits on the integral over θ are 0 and $\pi/2$. Division of the result by dS dt yields

$$\frac{\mathrm{d}N'}{N} = \pi\left(\frac{m}{2\pi kT}\right)^{3/2}e^{-mu^2/2kT}u^3\,\mathrm{d}u \tag{3}$$

for the fraction of molecules per second with speeds in the interval du that strike a unit area on the left side of the wall. The integral over u yields the factor $2(kT/m)^2$ and Eq. (3) leads to the result

$$N' = N\sqrt{\frac{kT}{2\pi m}} = \tfrac{1}{4}N\bar{u}, \tag{4}$$

where $\bar{u} = \sqrt{8kT/\pi m} = \sqrt{8RT/\pi M}$, as given by Eq. (I.59). In conclusion, Eq. (4) shows that the rate of molecular collisions with a unit area on the wall is one quarter of the average molecular speed.

Now suppose that a small hole of diameter d is cut in the wall and that a vacuum is created on the right side (r). If the gas pressure is sufficiently low and the hole is small enough, then the mean-free path will be large with respect to the hole diameter, i.e. $\lambda_o \gg d$. Under these conditions a given molecule will pass through the hole without colliding with another molecule, or the wall. Here, it is assumed that the wall is thin with respect to d. The number of such molecules arriving on the vacuum side of the wall will then be given by Eq. (4), The type of gas flow described here is referred to as effusive flow and the process is called effusion.

From Eq. (4) it is found that the volume of effusing gas per second per unit area of the orifice is equal to $\sqrt{RT/2\pi M}$. Hence, at a given temperature it should vary inversely with the square-root of the molecular weight. This law of effusion is attributed to Graham, whose early experiments verified it (see Table 1 and Partington in 'Further Reading' at the end of the book).

However, the conditions for effusive flow are not met when, at ordinary pressures and with a macroscopic hole size, a molecule cannot pass without undergoing collisions with other molecules. The net effect is to develop a collective-flow behavior and, of course, Eq. (4) is not applicable. When $\lambda_o \ll d$, the correct treatment of the problem is carried out with the use of the theory of viscous flow. This subject will not be presented here.

In the above analysis of effusive flow it was assumed that the gas passed through an orifice into a vacuum. However, if the wall separates two containers that are at different pressures, both of which are so low that $\lambda_o \gg d$, then the molecules from each container will flow into the other. If both containers are at the same temperature, then the flow rate given by Eq. (4) can be written in the form

$$f = \tfrac{1}{4}(n_\ell - n_r)\bar{u} = \frac{\bar{u}}{4kT}(P_\ell - P_r) \tag{5}$$

for an ideal gas, where n represents the number of molecules per unit volume and the subscripts ℓ and r refer to the left and right sides of the wall, respectively (see Fig. 1).

Table 1   Effusion of gases

Gas	Relative speed of effusion	
	Observed	Calculated [Eq. (4)]
H_2	3.623	3.802
CH_4	1.328	1.342
N_2	1.014	1.014
C_2H_4	1.013	1.014
Air	1[a]	1[a]
O_2	0.950	0.952
CO_2	0.831	0.808

[a] Air was used as a reference.

For the case of a mixture of ideal gases, Eq. (5) can be applied separately to each component, as no collisions occur near the orifice. In a two-component system, for example, the ratio of the flow rates is given by

$$\frac{f_1}{f_2} = \frac{\bar{u}_1}{\bar{u}_2}\frac{P_{1\ell} - P_{1r}}{P_{2\ell} - P_{2r}} = \left(\frac{M_2}{M_1}\right)^{1/2}\frac{P_{1\ell} - P_{1r}}{P_{2\ell} - P_{2r}}. \tag{6}$$

If the partial pressures on the right-hand side are maintained much lower than those on the left, then the effusion process results in a change in the composition of the mixture such that

$$\frac{n_{1r}}{n_{2r}} \approx \left(\frac{M_2}{M_1}\right)^{1/2}\frac{n_{1\ell}}{n_{2\ell}}. \tag{7}$$

Thus, if the two gases have different molecular weights, then the mixture effusing on the right-hand side will become richer in the lighter component. The square root of the mass ratio in Eq. (7) represents the upper limit to the separation factor, as in practice the partial pressures on the right cannot be completely neglected.

The effusion method is often used to enrich or separate the components of a gaseous mixture. Usually, permeable barriers with very fine pores are employed. However, because the lengths of the pores are considerably greater than their diameters, the gas flow does not follow the simple effusion analysis outlined above. Nevertheless, the molecular weight dependence is the same as that given by Eq. (7), because each molecule passes through the barrier independently of the others. This method is of particular importance in isotopic separation, such as that of H_2 and D_2—an isotopic pair for which the separation factor is particularly favorable. However, the separation of, say, $^{235}UF_6$ and $^{238}UF_6$ is a much less efficient process!

Chapter 10

Chemical Reactions[*]

10.1 RATE OF REACTION

In general, the rate of a chemical reaction in the gas phase can be expressed as a function of the concentration (indicated by the square brackets) or the partial pressure of each constituant. Thus,

$$\frac{\mathrm{d}[A]}{\mathrm{d}t} = f\{[A], [B], [C], \ldots [N]\}. \tag{8}$$

The rate of formation of species A is then a function of the concentrations of all the reactants and products involved, A, B, $\ldots N$. In the classical formulation of the kinetics of chemical reactions, the rate of disappearance of species A is written in the form

$$-\frac{\mathrm{d}[A]}{\mathrm{d}t} = k[A]^a[B]^b[C]^c \ldots [N]^n. \tag{9}$$

The overall order of the reaction is then given by the sum of the exponents, namely $a + b + c + \ldots + n$. A given exponent in Eq. (9) specifies the order of the reaction with respect to the species in question. The exponents might be expected to be positive integers (or zero), but such is not always the case. Reactions of fractional order, and even negative order, are known.

A particularly simple case is that of a first-order reaction with respect only to a given species, say, A. Then Eq. (9) becomes

$$-\frac{\mathrm{d}[A]}{\mathrm{d}t} = k[A]. \tag{10}$$

The integration of this expression yields

$$\ln \frac{[A]}{[A_0]} = -kt, \tag{11}$$

where $[A_0]$ is the initial concentration of species A. It is evident that a logarith-

[*]See Laidler in 'Further Reading' at the end of the book.

mic plot of the concentration of A as a function of time should be linear with a slope $-k$. This test is applied to establish that the reaction is indeed of first order.

An example of a simple first-order chemical reaction is the decomposition of ethyl bromide according to the reaction

$$C_2H_5Br = C_2H_4 + HBr. \tag{12}$$

Other processes that obey the first-order rate law include the radiation of light from an excited atom or molecule, as well as radioactive alpha and beta decays. A radioactive decay is usually characterized by its half-life, which can be easily related to the rate constant defined in Eq. (11). If the half-life is given by $t = t_{1/2}$ at $[A] = [A_0]/2$,

$$t_{1/2} = \frac{1}{k} \ln 2. \tag{13}$$

Furthermore, it should be noted that the rate constant k has dimensions of reciprocal time. Therefore, a time that is characteristic of the reaction can be defined by $\tau = 1/k$, and Eq. (13) becomes

$$t_{1/2} = \tau \ln 2. \tag{14}$$

The quantity $\tau_{1/2}$ is then the time for the concentration or partial pressure of the reactant to decrease by a factor of $1/e$ [see Eq. (11)].

It should be emphasized here that, in general, the measured order is not simply related to the mechanism of the reaction. The order is determined from experimental data, with the aid of the various rate laws. The mechanism of the reaction is a model suggested to relate the order to the so-called molecularity, a quantity that specifies the number of atoms or molecules involved in a reactional collision. Hence, a unimolecular reaction is a simple decomposition, as in the example of Eq. (12). A bimolecular reaction results from the collision of two reactants. The latter does not necessarily result in a second-order reaction!

An interesting case is that of a reaction of zero order. In this case, the rate is independent of the concentrations of all species present. A photo-chemical reaction in which all of the incident electromagnetic radiation is absorbed by the reacting molecules is an example of a reaction of zero order. Another is a reaction taking place on a catalytic surface at a molecular concentration that is sufficient to cover the entire surface.

10.2   ENERGY OF ACTIVATION

The concept of an energy of activation is usually attributed to Arrhenius (1889). To account for the temperature dependence of rate constants, he suggested that an equilibrium exists between 'inert' and 'active' molecules of the reacting species. Only the latter were presumed to be able to enter into reaction. Then, if

E_a represents the energy difference between the two kinds of molecule, the rate constant can be written in the form

$$k = A\,e^{-E_a/RT}, \tag{15}$$

as suggested by the Boltzmann factor, which was introduced in Chapter 2.

The pre-exponential factor A was initally assumed to be temperature-independent, although, as indicated below, both theoretical and experimental evidence suggests at least a weak temperature effect. If, for the present, it is assumed that both A and E_a are independent of temperature, measurements of the rate constant for a given reaction as a function of temperature should provide the necessary data for the determination of both A and E_a. From Eq. (15) it is apparent that a plot of $\ln k$ vs. $1/T$ would yield a straight line with slope $-E_a/R$ and intercept $\ln A$.

The exponential factor in Eq. (15) represents the fraction of molecules that possess the requisite energy to enter into reaction. The quantity E_a is, therefore, referred to as the energy of activation. As the pre-exponential factor A has dimensions of reciprocal time, it is often called the frequency factor.

10.3   A COLLISION MODEL

Perhaps the simplest chemical reaction could result from the collision of two atoms to form a diatomic molecule. This case corresponds to a bimolecular reaction, as described above. The number per second of binary collisions of two different atoms can be obtained by modifying Eq. (I.63), which then takes the form

$$Z_{ab} = 4n_a n_b \sigma_{ab}^2 \sqrt{\frac{\pi kT}{2\mu_{ab}}}. \tag{16}$$

In Eq. (16), $\sigma_{ab} = (\sigma_a + \sigma_b)/2$ is the mean diameter of the colliding atoms, and μ_{ab} is their reduced mass [see Eq. (I.186)]. If Z_{ab} is identified with the pre-exponential factor in Eq. (15), the Arrhenius equation becomes

$$k = Z_{ab}\,e^{-E_a/RT}. \tag{17}$$

Since A is now proportional to $\sqrt{T}$, Eq. (15) leads to the relation

$$\ln k = (B + \tfrac{1}{2}\ln T) - \frac{E_a}{RT}, \tag{18}$$

where B is a constant. Thus, the plot based on Eq. (15), as suggested above, is no longer linear, although the temperature dependence of A may not be apparent over a limited range of experimental temperatures. This simple collision model has been applied with some success to bimolecular reactions involving atoms and/or simple molecules.

Since the collision frequency Z_{ab} can be determined with the use of molecular diameters obtained from viscosity data (see Chapter 5), E_a can be measured from the temperature dependence of the rate constant for a given reaction. These experiments have been carried out for a variety of bimolecular reactions in the gas phase. Gaseous reactions involving simple molecules, such as $H_2 + I_2 = 2HI$, are in reasonable agreement with the predictions of Eq. (17). However, in many other cases, reactions have been found to be slower. To 'explain' this phenomenon, a so-called steric factor p was introduced. Then, Eq. (17) is written in the form

$$k = pZ_{ab}\, e^{-E_a/RT}. \tag{19}$$

The factor p, however, must take on very small values in certain cases (10^{-1} to 10^{-8}), a result that is difficult to justify on any obvious physical grounds.

It was pointed out long ago that, by thermodynamic arguments, the activation energy E_a should be replaced by the change in free energy of the reaction, namely $F_a = H_a - TS_a$. Thus, Eq. (15) becomes

$$k = Z_{ab}\, e^{S_a/R}\, e^{-H_a/RT} \tag{20}$$

and the steric factor can then be directly interpreted in terms of the entropy of activation, at least in the case of bimolecular reactions. However, the extension of this theory to the reaction of more complex molecules becomes doubtful, as it is difficult to take into consideration the role of the various internal degrees of freedom of the reacting species.

10.4   THEORY OF ABSOLUTE REACTION RATES

A more general theoretical approach to the interpretation of kinetic data employs statistical mechanics. It is known as the theory of absolute reaction rates [1]. It is based on a model in which a chemical reaction or other rate process is characterized by an initial ensemble of reacting species that pass via continuous coordinate changes into a final configuration—that which describes the products of the process. There is, however, along the route an intermediate state. This state is critical in the sense that, if it is attained, there is a high probability that the reaction will continue to completion. The intermediate species is known as the activated complex or, sometimes, the transition state. It includes all internuclear configurations of a reacting system that are intermediate between reagents and products. It is fundamental to the understanding of the processes of chemical change. As a simple example, consider a reaction of the type $A + BC \rightarrow (ABC)^{\ddagger} \rightarrow AB + C$. In this case, the activated complex is represented by $(ABC)^{\ddagger}$.

In the initial formulation of the theory of absolute reaction rates the activated complex was hypothetical in nature. More recent experimental studies with the use of time-resolved spectroscopy have, in a number of cases, identified this intermediate species. This subject will be briefly summarized in Section 10.5. As the activated complex is an ordinary—albeit short-lived—molecule, it possesses all the usual thermodynamic properties. However, one of its translational degrees of freedom is reserved to describe its displacement along a reaction coordinate, its path towards its decomposition at a definite rate.

For the statistical treatment of reaction rates it is assumed that the reactants are always in equilibrium with the activated complex. The latter can be imagined to exist at the top of an energy barrier, as suggested by Fig. 4. The rate at which the reaction occurs is determined by the average velocity of the system in the forward direction over the top of the barrier.

The transition state represents the existence of the activated complex within a small distance δ at the summit of the potential barrier (Fig. 4). The average speed of displacement of this species along the reaction path r is then given by

$$\bar{r} = \sqrt{\frac{kT}{2\pi m^{\ddagger}}}, \tag{21}$$

where $m^{\ddagger}$ is the mass of the activated complex. The mean lifetime of the activated complex, which corresponds to the average time of crossing the barrier is then

$$\frac{\delta}{\bar{r}} = \delta\sqrt{\frac{2\pi m^{\ddagger}}{kT}}. \tag{22}$$

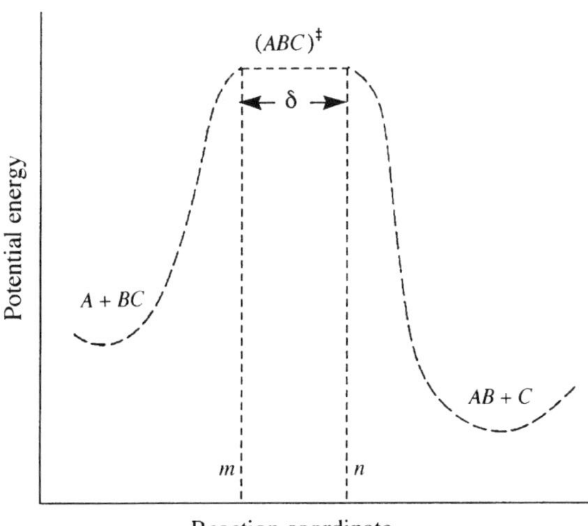

Fig. 4 Schematic diagram of the variation of the potential energy for the reaction $A + BC \rightarrow AB + C$ (redrawn from Fig. 9A–2 in *Molecular Theory of Gases and Liquids* by J. G. Hirschfelder, C. F. Curtiss and R. B. Bird, by permission of John Wiley & Sons, Inc., New York, 1954)

If $[(ABC)^{\ddagger}]$ is the number of activated complexes per unit volume within the interval δ, the rate of reaction is the number of complexes that cross the barrier per unit volume in unit time, namely $[(ABC)^{\ddagger}]/[\delta\sqrt{kT/2\pi m^{\ddagger}}]$. This quantity can be equated to the rate of reaction, which is given by $k[A][BC]$. Then, the rate constant for the reaction becomes equal to

$$k = \frac{[(ABC)^{\ddagger}]}{[A][BC]} \sqrt{\frac{kT}{2\pi m^{\ddagger}}} \frac{1}{\delta}. \tag{23}$$

It is assumed here that every complex that moves across the barrier continues to reaction. However, it is sometimes necessary to include a somewhat arbitrary factor, the transmission coefficient, to adjust this result to fit experimental data.

For a simple bimolecular reaction involving species A and BC, the equilibrium constant that relates the reactants to the activated complex $(ABC)^{\ddagger}$ is given by

$$K^{\ddagger} = \frac{[(ABC)^{\ddagger}]}{[A][BC]}. \tag{24}$$

From the definition of the partition function (see Chapter 3) it is apparent that it is proportional to the number of molecules in a given volume. Thus, the equilibrium constant expressed by Eq. (24) can be written in the form

$$K^{\ddagger} = \frac{\mathscr{Z}_{(ABC)^{\ddagger}}}{\mathscr{Z}_A \mathscr{Z}_{BC}}. \tag{25}$$

The partition function for the activated complex will include the rotational and vibrational degrees of freedom of a diatomic molecule, as well as the three degrees of translational motion. One of the translational degrees of freedom is identified with the movement of the complex along the minimal reaction path on the potential energy surface. Then, from Eq. (I.76)

$$\mathscr{Z}_{(ABC)^{\ddagger}} = \mathscr{Z}'_{(ABC)^{\ddagger}} \frac{\sqrt{2\pi m^{\ddagger} kT}}{h}, \tag{26}$$

where $\mathscr{Z}'_{(ABC)^{\ddagger}}$ includes the remaining degrees of freedom of the activated complex. Substituting δ from Eq. (23) and defining the equilibrium constant K for the reaction by

$$K = \frac{[AB][C]}{[A][BC]}, \tag{27}$$

the rate constant is then given by

$$k = \frac{kT}{h} K. \tag{28}$$

It should be noted that δ, the rather arbitrary distance along the reaction path, disappears in this derivation. Thus, the rate constant for the reaction, as given in

Eq. (28), contains the very important temperature-dependent factor kT/h. This factor is of the order of magnitude of 10^{13} s^{-1} at room temperature.

The above analysis allows an interpretation in terms of a certain reaction path between the reactants and the products, passing by the existence of the activated complex in the region of a saddle point. A schematic diagram of this process is shown in Fig. 5, where a hypothetical potential surface between reactants and products has been introduced. While this picture of a chemical reaction may seem somewhat abstract, the potential surface has been calculated for very simple reactions such as $D + H_2 \rightarrow HD + H$ [2]. However, the confirmation of these predictions had to await the development of much more sophisticated experiments (see Zewail in 'Further Reading' at the end of the book). The example that is described in Chapter 12 employs crossed molecular beams to obtain specific information concerning such chemical reactions.

The rate constant given by Eq. (28) does not provide a precise molecular description of a chemical reaction because it represents an average over all of the possible encounters of the reacting species. Thus, the mechanism of a collision between reactants includes, in general, their relative velocities, mutual orientations, vibrational and rotational motions and impact parameters (see Chapter 8). What is needed here is a specific description of a chemical reaction: how molecules collide, exchange energy, break and/or form bonds, before their separation into products. In this Chapter an attempt will be made to describe experimental investigations of these questions.

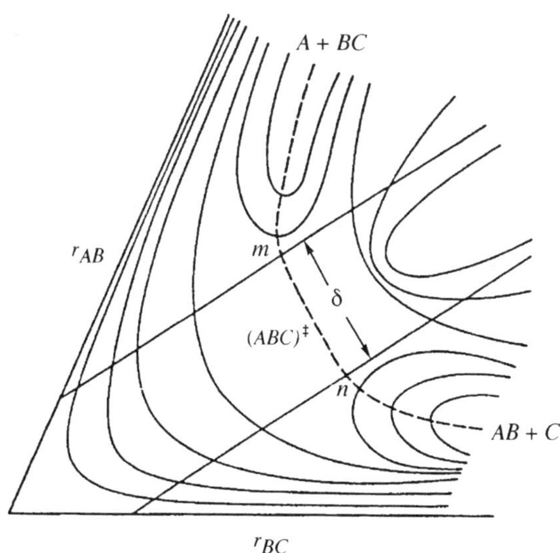

Fig. 5 Potential energy surface for the reaction $A + BC \rightarrow AB + C$ (redrawn from Fig. 9A–1 in *Molecular Theory of Gases and Liquids* by J. G. Hirschfelder, C. F. Curtiss and R. B. Bird, by permission of John Wiley & Sons, Inc., New York, 1954)

10.5   REAL-TIME OBSERVATIONS

The frequency factor, kT/h, in Eq. (28) corresponds to a time scale of the order of 10^{-13} s, or 0.1 ps (picosecond). This result discouraged early researchers from having any hope of detecting directly the intermediates in chemical reactions. It was felt that there was no possibility of observing such short-lived species. However, some early attempts to follow in real time the evolution of chemical reactions were made with the use of the flash-photolysis technique. These experiments employed a flash lamp to produce a pulse of light that initiated a chemical change, followed by a probe or spectroscopic flash used as a source to record the situation spectroscopically a short time later. A simplified diagram of the apparatus is shown in Fig. 6. Initially, the time scale of these observations was of the order of milliseconds; although, with refinement of the technique, times as short as a few microseconds were obtained.

The flash-photolysis method was used in the 1950s to study a number of relatively simple chemical reactions [3]. Among them, a particular interesting one was the reaction of oxygen and chlorine. The principal steps involved are:

(i) $$Cl_2 + h\nu \Rightarrow 2Cl,$$
(ii) $$2Cl + O_2 \Rightarrow 2ClO,$$
(iii) $$2ClO \Rightarrow Cl_2 + O_2.$$

The first is the dissociation of chlorine, which was triggered by the arrival of a short pulse of ultraviolet light from the flash lamp. A millisecond or so later, with the arrival of the second pulse, the absorption spectrum of the gaseous system was recorded. The free radical ClO resulting from reaction (ii) was identified by its ultraviolet spectrum, while its lifetime of a few milliseconds was measured by successively varying the time between the two pulses. These experiments represent the birth of what is now often referred to as 'pulse-and-probe spectroscopy'.

It is interesting to note that the radical ClO results from the destruction of oxygen molecules by chlorine atoms. Later experiments revealed that ozone, O_3, is also attacked by chlorine atoms to produce ClO. This observation became

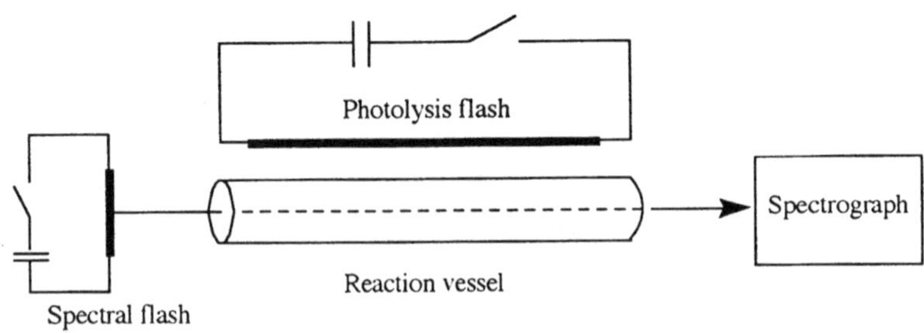

Fig. 6   Diagram of the apparatus for a flash-photolysis experiment

important much more recently, when concern arose over the depletion of the earth's ozone layer. It has now been proven that chlorine atoms produced by the photodecomposition of freon molecules attack the ozone in the stratosphere. This process is seasonally dependent, as shown by direct measurements of both ozone and ClO concentrations over Antarctica at various times of the year [4, 5].

Pulse-and-probe spectroscopy has evolved rapidly since the invention of the laser in the early 1960s. Its progress can be measured by the time resolution of the measurements. As very short laser pulses became available, a split-beam method was developed [6], as shown in Fig. 7. A laser source produces a very short pulse of light that is directed toward a beam splitter. This optical element plays the role of a mirror, as well as a window. Thus, say, half of the light is transmitted as beam (*a*), while the other half is reflected as beam (*b*). Beam (*b*) is deflected toward a corner mirror and returns to the beam splitter, where, by reflection, it is combined with beam (*a*). Clearly, beam (*b*) is delayed in time with respect to beam (*a*). The delay time is determined by the distance d indicated in Fig. 7. Thus, the second pulse, which serves to analyze the system, arrives at a precisely determined time after the arrival of the first one. This method assures an exact time relation between the two pulses. It serves as the basis for virtually all current real-time measurements of chemically reacting systems. At the present time, kinetic studies are being carried out on the femtosecond time scale (1 fs $= 10^{-3}$ ps $= 10^{-15}$ s), or at least at some tens of femtoseconds. However, the basic principle has not changed, even if the advances in laser technology are impressive.

As an example of this work, consider the photo-induced dissociation of ICN [7]. The reaction can be represented by

$$ICN + h\nu \Rightarrow ICN^* \Rightarrow [I \ldots CN] \Rightarrow I + CN,$$

where the photon $h\nu$ corresponds to light at a wavelength $\lambda_1 = 307$ nm. From the potential energy diagram of Fig. 8, it is apparent that the absorption of a short pulse at this wavelength will excite the molecule to the repulsive level V_1, resulting in the production of ICN* with an IC bond length equal to r_e. The

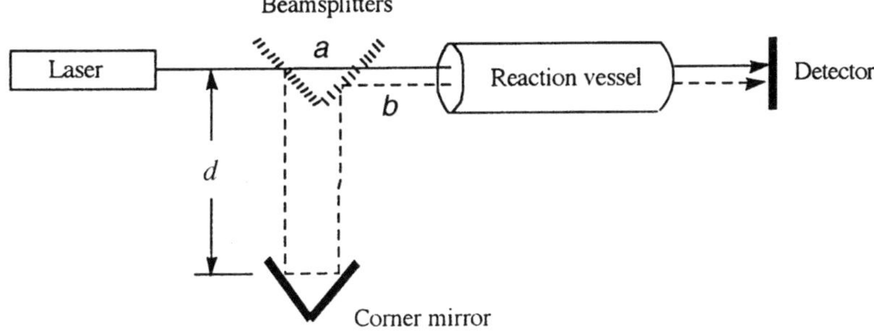

Fig. 7 Diagram of a pump-probe experiment

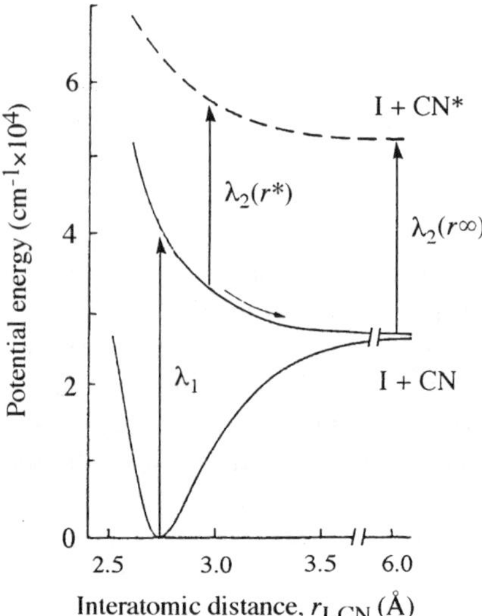

Fig. 8 Potential energy diagram for the dissociation of ICN (reprinted with permission from M. Dantus, M. J. Rosker and A. H. Zewail, *J. Chem. Phys.* **87**, 2395 (1987). Copyright (1987) American Institute of Physics.)

excited molecule then begins to dissociate, forming the intermediate species [I ... CN], in which the distance I ... C increases monotonically with time. As this distance approaches infinity, the reaction ends with the formation of the separated products. The concentrations of either the final product CN, or the transient species, can be followed with the use of an appropriate probe pulse. The wavelength of the probe pulse, which is delayed by varying time intervals with respect to the pump pulse, is adjusted to induce fluorescence of the CN product. The intensity of the fluorescence resulting from a probe at $\lambda = 388.9$ nm exhibits a build-up to a constant value a few hundred femtoseconds later, as shown in Fig. 9(a). An adjustment of the probe wavelength to 389.5 nm results in the monitoring of the transient species, because the I ... C distance increases from its equilibrium value in the ICN molecule. Curve (b) of Fig. 9 increases with time to a maximum before decreasing, because the probe pulse is no longer in resonance. A pictorial representation of this photofragmentation reaction is presented as Fig. 10 [8].

The two examples of real-time studies of simple chemical reactions have been chosen to illustrate the evolution of the method over the past 30 years, or so. The technique is now being applied to many different chemical systems and promises to be extremely useful in the investigation of biomolecular reactions. Its combination with other experimental methods, such as molecular beams (see Chapter 11), has already been demonstrated. It might be asked whether a continued improvement in time resolution can be achieved. It probably can, but, as pointed

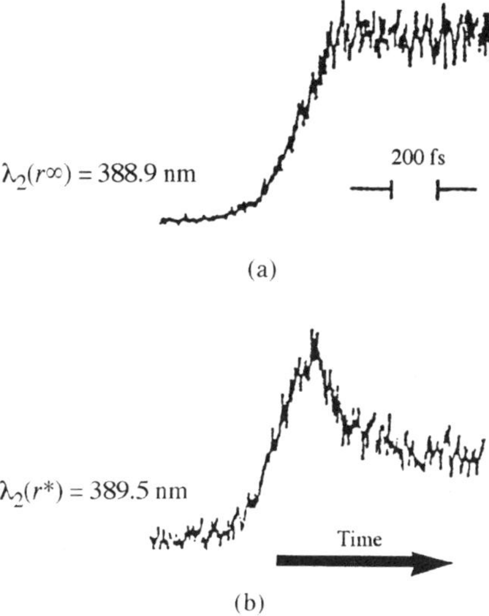

Fig. 9 Time dependence of fluorescence intensities during the photolysis of ICN (reprinted with permission from M. Dantus, M. J. Rosker and A. H. Zewail, *J. Chem. Phys.* **87**, 2395 (1987). Copyright (1987) American Institute of Physics, New York, NY)

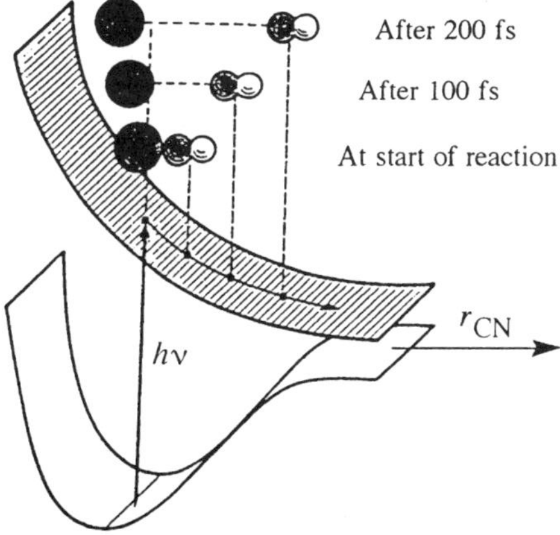

Fig. 10 Photofragmentation of ICN (adapted with permission from M. Dantus, M. J. Rosker and A. H. Zewail, *Science* **241**, 1200 (1988). Copyright (1988) American Association for the Advancement of Science)

out above, the pre-exponential factor in Eq. (15) is of the order of magnitude of 100 fs, the minimum value for the lifetime of the transition state. Thus, the tens-of-femtoseconds resolution now available is quite sufficient for real-time measurements of even the fastest chemical reactions.

A second, very fundamental argument is based on Heisenberg's uncertainty principle, according to which it is usually stated that the position and momentum (in the same direction) of a particle cannot both be determined with precision at the same time. An analogous uncertainty relation exists between the energy and the time. Since spectroscopic transitions involve changes in energy, and thus in frequency ($\Delta E = h\nu$, according to Planck), the uncertainty in the time of a transition and that of its frequency are similarly related. The result is that virtually all spectroscopic information about molecules (the movements of nuclei) becomes lost at times of a femtosecond or less. This question will be considered further in connection with the broadening of spectral lines (see Chapter 14).

Chapter 11

Ortho and *Para* Hydrogen

In the analysis of molecular energies presented in Chapter 4 it was assumed that the total energy was simply the sum of the translational, rotational and vibrational contributions. Thus, the partition function for a molecule was written as the corresponding product [see Eq. (I.112)]. The electronic part of the partition function was taken equal to one, which is correct for molecules at ordinary temperatures in a non-degenerate ground state (see, however, Chapter 16).

A particular problem arises in the case of molecules possessing a center of inversion, the simplest example being a homonuclear diatomic molecule. The symmetry can then be specified by the result of exchanging (permuting) the two identical and indistinguishable nuclei. From a quantum-mechanical point of view, the effect of the permutation on the wave function for the nuclear spins is either symmetric (without change in sign) or antisymmetric (with change in sign). If the wave function is symmetric, the molecule is called *ortho* and if it is antisymmetric, it is called *para*. It can be shown that the number of independent spin wave functions (spin degeneracy or statistical weight) is given by $g_i(ortho) = (i + 1)(2i + 1)$ if the wave function is symmetric under the permutation and $g_i(para) = i(2i + 1)$ if it is antisymmetric. Here, i is the spin of one of the identical nuclei. It has integer values or half-integer values, depending on the nature of the nucleus. In the former case, it is called a Boson, while in the latter it is known as a Fermion. For example, the proton has a spin of one-half and is thus a Fermion.

The general description of a homonuclear diatomic molecule must include the symmetry argument presented in the above paragraph, as well as the symmetries of the translational, rotational and vibrational wave functions. In Chapter 4 the energies associated with these degrees of freedom were given, although the wave functions were not. It is sufficient here to point out that both the translational and vibrational wave functions are symmetric under permutation of the nuclei. However, the symmetry of the rotational wave function depends on the values of the rotational quantum number, J. It is found that the wave function is symmetric if J is even and antisymmetric if J is odd. The overall symmetry of a homonuclear diatomic molecule is then determined by the product of the rotational and nuclear-spin wave functions. This analysis is employed in the application of Pauli's principle; namely, that the total wave function must be antisymmetric under the permutation operation.

11.1   STATISTICAL THERMODYNAMICS

A gas composed of homonuclear diatomic molecules must be considered to be a mixture of two different sorts, *ortho* and *para*, depending on the symmetry of the nuclear-spin wave function. To assure that the overall wave function is antisymmetric, the rotational wave function must be correctly chosen. Thus, for the *ortho* species, the rotational wave function is antisymmetric, corresponding to odd values of J in the expression for the partition function. Clearly, only even values of J are involved in the partition function for the *para* molecules. The specific forms of the partition functions are given by

(i)   *ortho* molecules:

$$\mathscr{Z}_{\mathrm{ns}}\mathscr{Z}_{\mathrm{rotation}}(J \text{ odd}) = (i+1)(2i+1)\sum_{J \text{ odd}}(2J+1)\,e^{-J(J+1)\theta_{\mathrm{rot}}/T} \qquad (29)$$

and

(ii)   *para* molecules:

$$\mathscr{Z}_{\mathrm{ns}}\mathscr{Z}_{\mathrm{rotation}}(J \text{ even}) = i(2i+1)\sum_{J \text{ even}}(2J+1)\,e^{-J(J+1)\theta_{\mathrm{rot}}/T}, \qquad (30)$$

where $\mathscr{Z}_{\mathrm{ns}}$ is the partition function for nuclear spin and θ_{rot} is the rotational temperature defined in Chapter 4. It should be noted from Eq. (29) that the energy of the ground state ($J=1$) is not, in this case, equal to zero.

The hydrogen molecule will now be taken as an example [9]. With $i = 1/2$ for the proton, the statistical weights become $(i+1)(2i+1) = 3$ and $i(2i+1) = 1$ in Eqs (29) and (30), respectively. Since the translational and vibrational partition functions for the two types of hydrogen are identical, the ratio of Eq. (29) to Eq. (30) is the *ortho*/*para* pressure ratio. This ratio approaches a value of three at high temperatures ($T \gg \theta_{\mathrm{rotation}}$) because the two sums become identical, each being given by one-half the value expressed by Eq. (I.86). In the low-temperature limit ($T = 0$ K), the sum in Eq. (29) becomes equal to zero and at equilibrium only *para*-H_2 can exist. This result, as expressed by the percentages of the two types of molecular hydrogen, are presented as a function of temperature in Fig. 11. However, as indicated in what follows, the equilibrium between these species is only very slowly established.

The evaluation of the various thermodynamic quantities from the partition functions is direct; although, for the rotational contribution, the partition function must be considered to be the weighted sum of those for the *ortho* and *para*

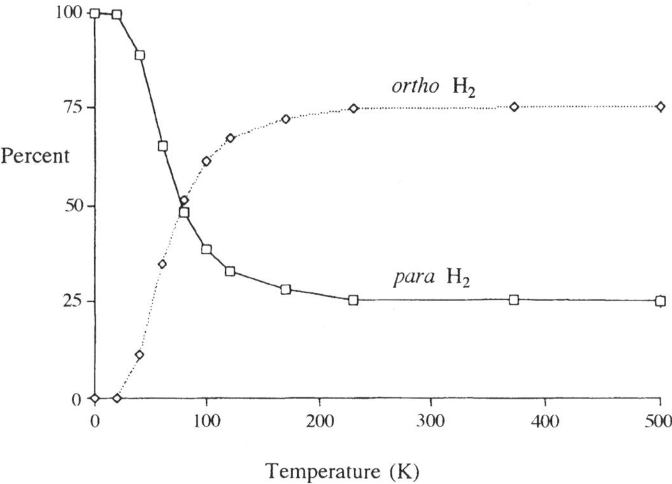

Fig. 11 Relative equilibrium percentages of *ortho* and *para* hydrogen as a function of temperature

forms. For example, the rotational part of the heat capacity is calculated from Eq. (I.45) with $\mathscr{Z}_{\text{rotation}} = \mathscr{Z}_{\text{rotation}}(ortho) + \mathscr{Z}_{\text{rotation}}(para)$. At ordinary temperatures, the equilibrium constant for the reaction *ortho*-$H_2 \leftrightarrow$ *para*-H_2 has almost reached its limiting value of $1/3$, as determined by the quantity $i/(i+1)$. At lower temperatures, the sums involved in the rotational partition functions must be evaluated term-by-term. The calculated rotational heat capacity is shown as a function of temperature in Fig. 12 for both *ortho*-H_2 and *para*-H_2, as well as for the equilibrium mixture.

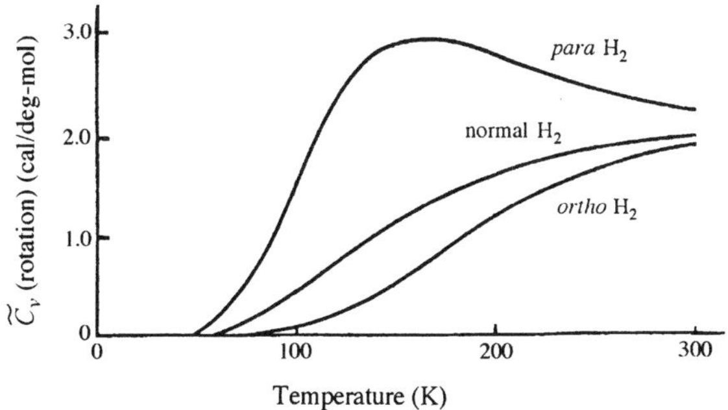

Fig. 12 Rotational contribution to the heat capacity $\tilde{C}_V$ for *ortho*, *para* and normal hydrogen

11.2 EQUILIBRIUM

As indicated above, in those changes in which only the spins of the nuclei are altered, the translational and vibrational contributions to the partition function are identical. Thus, for the reaction *ortho*-$H_2 \Rightarrow$ *para*-H_2, the pressure ratio at equilibrium is given by

$$K = \frac{[para\text{-}H_2]}{[ortho\text{-}H_2]} = \frac{\mathscr{Z}_{\text{rotation}}(para)}{\mathscr{Z}_{\text{rotation}}(ortho)}. \tag{31}$$

However, it should be emphasized here that equilibrium between the two types of hydrogen is not easily established. In fact, under ordinary conditions, equilibrium is achieved only after a period of some years. It was for this reason that the early experimental determinations of the heat capacity of ordinary hydrogen were not in agreement with the values predicted from the elementary theory, in which the symmetry considerations presented above were not taken into account. Furthermore, the spectra of molecular hydrogen obtained at that time displayed anomalous line intensities that suggested the presence of two sorts of hydrogen, as indicated in this simple theoretical analysis.

It was subsequently shown that the presence of a catalyst accelerates the attainment of the *ortho*−*para* equilibrium. Thus, in the presence of a suitable catalyst, a mixture of *ortho* and *para* hydrogen, usually with the latter in excess, is brought rapidly into equilibrium. Measurement of the properties such as heat capacity and thermal conductivity as a function of temperature then indicate that an equilibrium between the two species has been established.

It should be pointed out here that nuclear spins are not altered in the course of chemical reactions at ordinary temperatures. The statistical weight associated with nuclear spins then cancels out in expressions for the equilibrium constants of chemical reactions. However, in the present context, nuclear reactions—such as in an atomic bomb—have not been considered.

Chapter 12

Atomic and Molecular Beams

Atomic and molecular beams have been employed in a variety of important experiments in physics and chemistry. Some of these applications are described in this chapter.

To form a beam, a gas or vapor at relatively low pressure is usually allowed to effuse through a series of small holes or slits, as shown in Fig. 13. The collimated beam then passes through a selector that can choose a specific atomic or molecular velocity, or state, depending on the experiment.

12.1 DISTRIBUTION OF MOLECULAR SPEEDS

In the earlier experiments, a selector consisting of two rotating shutters was employed to let pass only those molecules with the desired velocity (Fig. 13). The shutters, which turn on the same shaft, are mounted with their slits angularly displaced by a few degrees. Thus, the speed of rotation of the shaft determines the velocity of the molecules that are detected, for example, by the formation of a deposit on a transparent screen. The number of molecules as a function of the speed of rotation of the velocity selector is measured from the density of the deposits. This device, which was originally suggested by Stern, was employed by

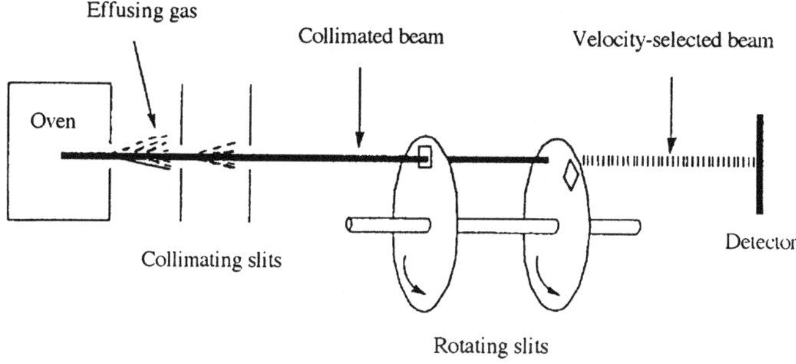

Fig. 13 Method of producing a velocity-selected molecular beam

several workers in the 1920s to study the Maxwell–Boltzmann distribution law (see Chapter 3).

It should be pointed out, however, that the velocity-selection system composed of two rotating shutters suffers from the fact that certain harmonics and, in particular, 'sub-harmonics' of the chosen velocity can pass to the detector. As an analogy, consider the passage of a series of cars along a route where the speed is controlled by traffic lights. If the lights are synchronized for the passage of cars at a certain maximum speed, then others can also pass if they move slowly enough to arrive for the next green light—and so on. Thus, a system was developed to avoid such artifacts in the detected velocity distribution. It consists of a rotating drum into which is cut a large number of helical slots, as shown in Fig. 14.

The rotating drum selector, which is very difficult to machine, has been employed in a number of experiments, including the velocity selection of slow neutrons. In the present application, Miller and Kusch [10] used a drum selector of this type for studies of the velocity distribution of several atomic species. Their analysis of this application is now illustrated and applied to the analysis of the velocity distribution in a beam of thalium atoms.

The velocity distribution in a molecular beam produced by gaseous effusion has been given by Eq. (3). The normalization of this expression to unity yields

$$N' = 2\left(\frac{m}{2kT}\right)^2 u^3\, \mathrm{e}^{-mu^2/2kT}. \tag{32}$$

The analysis of the flux of molecules falling on the entrance of the velocity selector shows that the intensity at the detector is given by $I = N'u\gamma$, where $\gamma = \ell/(r_o\phi_o)$ as shown in Fig. 14. It should be noted that this factor is characteristic of the velocity selector and is independent of the speed of its rotation. From Eq. (32) the intensity of the beam at the detector is then given by

$$I = 2\gamma\left(\frac{m}{2kT}\right)^2 u^4\, \mathrm{e}^{-mu^2/2kT}. \tag{33}$$

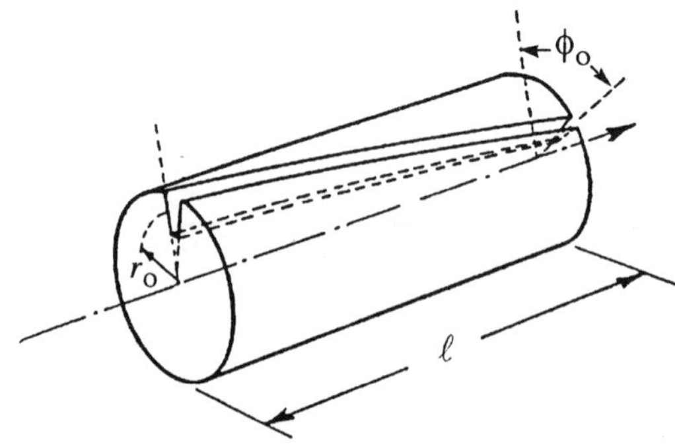

Fig. 14   Helical path velocity selector

Equation (33) has been applied in the analysis of experiments on potassium and thallium vapors [10]. The results of these very accurate measurements, as illustrated in Fig. 15 for thallium, provide a significant confirmation of the Maxwell–Boltzmann distribution law.

12.2 MOLECULAR DIFFRACTION

Molecular beams have been employed in direct measurements of collision cross-sections. A simplified diagram of the experimental setup is shown in Fig. 16. The collimated molecular beam enters a region containing a gas or vapor at low pressure that is composed of the same, or different, molecules as those of the beam. The collision of the beam molecules with the 'target' molecules obviously reduces the intensity of the beam reaching the detector. A simple analysis of this process is as follows.

If it is assumed that a velocity selector has been introduced in a perfectly collimated beam, then all of its molecules move at the same speed, u_0, in the direction of the beam as it enters the scattering region. As time goes on, the molecules in the beam will collide, one after the other, with target molecules and be scattered out of the beam. Assume now that the number of molecules originally in the beam at $t = 0$ is N_0 and at time t there are N remaining in the beam. In the following interval of time, dt, the number $Nk\,dt$ will undergo collisions and thus be scattered out of the beam. Here, k represents the collision

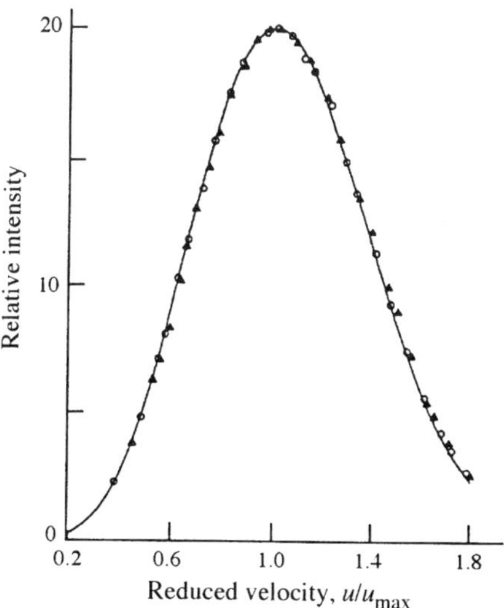

Fig. 15 Velocity distribution in a beam of thallium atoms: $\bigcirc$, $T = 870$ K; $\blacktriangle$, $T = 944$ K (reprinted with permission from Fig. 5 in R. C. Miller and P. Kusch, *Phys. Rev.* **99**, 1314 (1955). Copyright (1997) American Institute of Physics)

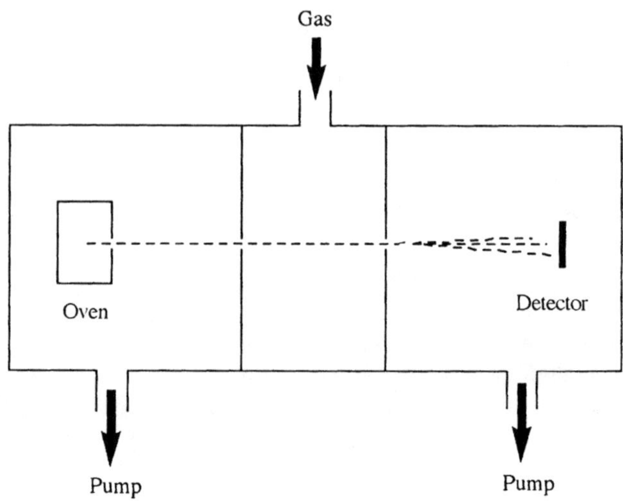

Fig. 16   Schematic representation of the apparatus for measuring molecular diffraction

rate for a beam molecule moving at speed u_0 among the target molecules that compose the gas. The corresponding change in N is then given by

$$\mathrm{d}N = -Nk\,\mathrm{d}t. \tag{34}$$

Division of Eq. (34) by N and integration yields the result

$$N = N_0\,e^{-kt} \tag{35}$$

$$= N_0\,e^{-k\lambda/u_0}, \tag{36}$$

where λ is the free-path distance traversed by a given beam molecule at time t. In these equations the initial condition $N = N_0$ at $t = 0$ has been imposed to evaluate the constant of integration. The similarity of Eq. (36) to Lambert's law in optics should be noted.

Suppose, now, that an experiment is made with the use of the apparatus shown in Fig. 16. First, the scattering region is evacuated and the number of molecules N_0 (the beam intensity) is measured by the detector. The scattering gas is then introduced and the measurement repeated to yield a value for N. Equation (36) is applicable, since only those molecules whose free path is equal to λ will reach the detector. Thus, Eq. (36) can be written in the form

$$N = N_0\,e^{-n_b Q\lambda}, \tag{37}$$

where an effective collision cross-section, $Q = k/(n_b u_0)$, has been introduced and n_b is the number of target molecules per unit volume. The cross-section can be measured directly in this type of experiment. If the molecules are considered to be hard spheres (Fig. I.16(a)), the cross-section is just $\pi\sigma^2$, as indicated in

Chapter 3. There, it was assumed that the beam molecules and the target molecules were of the same species. If they are different, the molecular diameter σ is replaced by $\sigma_{ab} = (\sigma_a + \sigma_b)/2$ [see Fig. I.7(a) and Eq. (I.142)].

Molecular diffraction experiments are often carried out without the use of a velocity selector. In this case, the velocity distribution is Maxwellian and the above analysis is applicable with u_0 replaced by $\bar{u}$, as a working approximation.

It is useful to introduce the general definition of the collision cross-section for classical binary encounters, namely

$$Q(g) = 2\pi \int_0^\infty (1 - \cos\chi) b \, db. \tag{38}$$

Here, b is the impact parameter defined in Chapter 8 and χ is the angle of deflection in classical binary collisions. The angle of deflection was derived previously [Eq. (I.203)]. The quantity $Q(g)$ is an effective cross-section for momentum transfer, which is a function of the initial relative velocity, g. The factor $1 - \cos\chi$ is maximum and equal to two for head-on collisions ($b = 0$) for which the momentum transfer is most effective. It becomes equal to zero as the impact parameter b approaches infinity, where $\chi = 0$ and no momentum is transferred.

To illustrate the physical significance of the collision cross-section, Eq. (38) will now be applied to the interaction of two hard spheres. Since the potential is equal to zero for $r < \sigma$, the angle of deflection is given in this case by

$$\chi = \pi - 2b \int_\sigma^\infty \left(1 - \frac{b^2}{r^2}\right)^{-1/2} \frac{1}{r^2} \, dr, \tag{39}$$

where σ is the molecular diameter for collisions between like molecules. The integration of Eq. (39) leads to

$$\chi = \pi - 2\sin^{-1}\left(-\frac{b}{\sigma}\right) \tag{40}$$

and

$$\sin\left(\frac{\chi - \pi}{2}\right) = -\frac{b}{\sigma}. \tag{41}$$

Therefore,

$$\frac{b}{\sigma} = \cos\frac{\chi}{2} = \left(\frac{1 + \cos\chi}{2}\right)^{1/2} \tag{42}$$

and

$$\cos\chi = \frac{2b^2}{\sigma^2} - 1. \tag{43}$$

Then, from Eq. (38), the cross-section becomes

$$Q(g) = 4\pi \int_0^{\sigma} \left(1 - \frac{b^2}{\sigma^2}\right) b \, \mathrm{d}b = \pi \sigma^2. \tag{44}$$

This result is, of course, what was introduced before for the collisions of like rigid spheres (see Chapter 3).

12.3   ELECTRON SPIN

The concept of a spinning electron evolved from observations of the splitting of lines in the spectra of atoms. A direct confirmation of this phenomenon, as well as a demonstration of spatial quantization, was provided by the experiments of Stern and Gerlach [11]. This work will be described briefly in order to illustrate the possibility of selecting specific atomic states with the use of a magnetic field.

When a magnetic dipole $\boldsymbol{\mu}$ is placed in a magnetic field $\mathbf{B}$, the energy of interaction is equal to $-\boldsymbol{\mu}\cdot\mathbf{B}$. Thus, if the field is inhomogeneous, the dipole is subjected to a force that depends on its orientation relative to the direction of the field. This force is given by the negative gradient of the interaction energy, namely

$$\mathbf{f} = \nabla(\boldsymbol{\mu}\cdot\mathbf{B}). \tag{45}$$

If the magnetic field is applied in, say, the z direction, then the force on the dipole is given by

$$f_z = \mu_z \left(\frac{\mathrm{d}B}{\mathrm{d}z}\right) = \mu \cos\theta \left(\frac{\mathrm{d}B}{\mathrm{d}z}\right), \tag{46}$$

where θ is the angle between $\boldsymbol{\mu}$ and $\mathbf{B}$. Thus, if a beam of atoms with a permanent magnetic moment, $\boldsymbol{\mu}$, passes through an inhomogeneous magnetic field, $\mathbf{B}$, then it will be deflected by an amount that depends on the relative orientations of these vectors, as well as their magnitudes—and the magnetic moment of the sample.

In the Stern–Gerlach experiment, a collimated beam of atoms effusing from an oven passes between the poles of a magnet, as shown in Fig. 17. The pole pieces are shaped so that the field has a strong gradient perpendicular to the direction of propagation of the beam. The condensation of the atoms on a screen allows the positions and relative intensities of the different components of the beam to be detected. In the absence of the magnetic field, the beam is focused on a single spot at $z = 0$ on the observation screen.

The atoms that effuse from the oven have their magnetic moments randomly oriented. Hence, from Eq. (46) the force f_z resulting from the magnetic field should vary continuously over the range $\pm\mu(\mathrm{d}B/\mathrm{d}z)$, corresponding to the limits $\theta = 0$ to $\theta = \pi$. A purely classical model of the interaction would then predict the deposit of a continuous band symmetrically distributed along the z axis. The

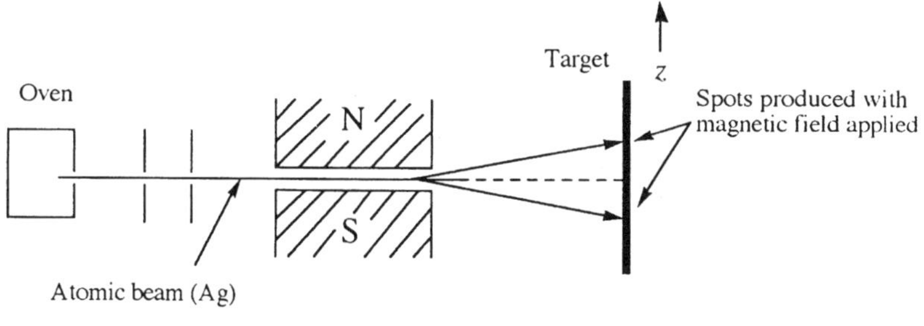

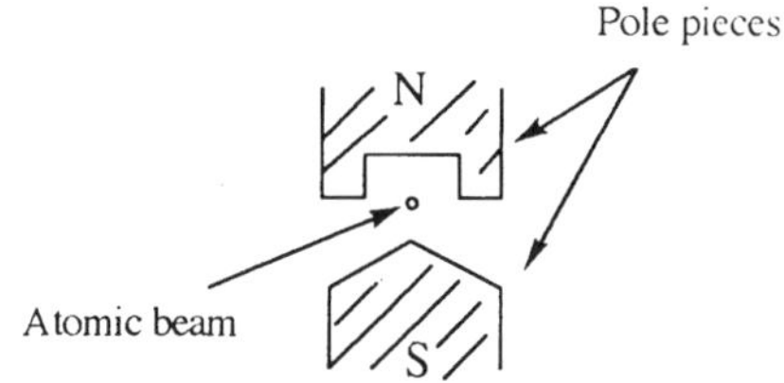

Fig. 17 The Stern–Gerlach experiment

fact that such a continuous pattern is never observed, but that the deposit is always in the form of a number of spots, provides direct evidence of the spatial quantization of the magnetic moments of atoms.

Consider a single-electron atom with total angular momentum $\mathscr{M}$. In classical electrodynamics, this quantity is directly related to the magnetic moment by

$$\boldsymbol{\mu} = \frac{-e}{2m_e} \mathscr{M}, \tag{47}$$

where $(-e)$ is the charge on the electron and m_e is its mass. Since, in quantum mechanics, the angular momentum is quantized, it is not surprising that the magnetic moment is also. As the angular-momentum quantum number can take on the values $\ell = 0, 1, 2 \ldots$, with multiplicity $2\ell + 1$, it would be reasonable to expect $2\ell + 1$ spots when the experiment is performed on the atom in a state s, p, d, ... as ℓ takes on the values 0, 1, 2, etc.

In the original experiment, a beam of silver atoms was produced by the evaporation of metallic silver in the oven. The silver atom in its ground electronic state has but a single $5s$ electron in its outer shell. Thus, $\ell = 0$ and the multiplicity is equal to one. Why, then, were two spots observed on the screen in this experiment?

To respond to the question posed in the above paragraph it was necessary to postulate the existence of another contribution to the total magnetic moment of the electron. It was called the 'spin', although the physical significance of this term might be questioned. Can a point charge, as the electron is assumed to be, have a spin? Without entering into the philosophy of this simple model—and that it is—consider its quantum-mechanical consequences.

It is assumed that the electron possesses an angular momentum $\mathscr{S}$ because of its spin. Furthermore, the component of $\mathscr{S}$ on the z axis can take on only two values, $\pm\hbar/2$. Thus, in the original experiment with silver atoms ($\ell = 0$) the two spots were produced by the equal and opposite contributions of these components. This result can be generalized to polyelectronic atoms, although it is necessary to consider the method of adding up the various angular-momentum components. It can be shown that when atoms with an odd number of electrons are employed in the Stern–Gerlach experiment, an even number of spots will always be produced. The analysis of this question is developed in most books on quantum mechanics.

12.4   THE AMMONIA MASER

The acronym 'MASER' was suggested by Townes and his group [12] in the 1950s to represent the expression 'microwave amplification by stimulated emission of radiation'. This term is now used to describe any device that amplifies signals or oscillates at microwave frequencies with the use of the quantum-mechanical principles of stimulated emission.

A few years after the development of the maser, analogous experiments were performed at optical frequencies. They resulted in the generation of coherent light beams and the 'LASER' (where 'microwave' is replaced by 'light' in the above acronym); the resulting devices are now everyday knowledge. Thus, it is important to present here a description of the ammonia maser, as originally introduced. Some of the more recently developed gas lasers will be discussed in Chapter 15.

The ammonia molecule, NH_3, has the structure of a pyramid whose equilateral-triangular base is defined by the three protons. This structure can be inverted in the sense that the protons can pass to equivalent positions on the other side of the nitrogen atom, as shown in Fig. 18. The relatively low barrier to this process (approximately 5.8 kcal/mol) results in splitting of the energy levels. The separation, $\varepsilon_a - \varepsilon_b$, between the resulting two components corresponds to a frequency of 23.87 GHz (≈ 0.79 cm^{-1}). Thus, an absorption line is readily observable at this frequency in the microwave spectrum of ammonia.

At equilibrium, the relative populations of the two levels of energies ε_a and ε are determined by the Boltzmann factor, $\exp\left[(\varepsilon_a - \varepsilon_b)/kT\right]$, as given in Chapter 2. Both types of molecule effuse into a vacuum from an oven whose temperature can be carefully regulated (Fig. 19). The resulting molecular beam then passes through an electric field produced by a system of four electrodes, as shown. This quadrupole field preferentially focuses the molecules in the upper state towards the center of the molecular beam. The selection of these molecules yields an excess of molecules in the upper state—the so-called population

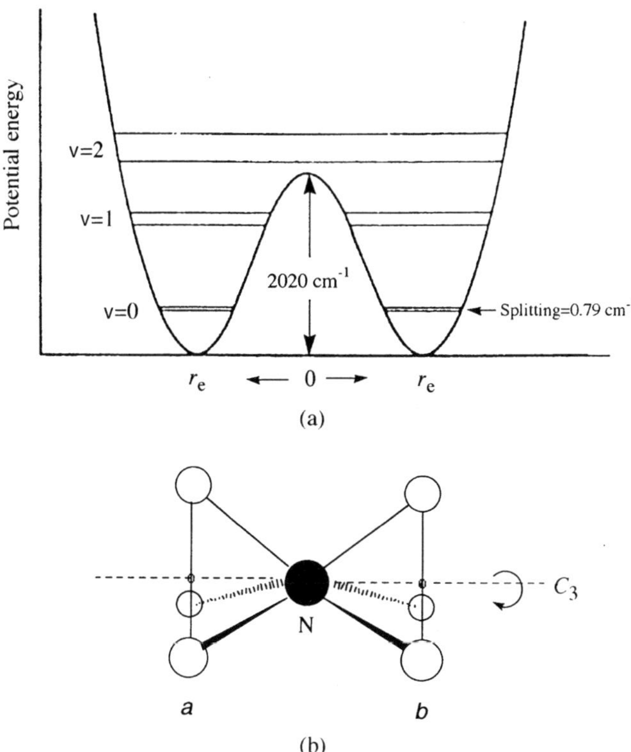

Fig. 18 (a) Potential function and energy levels for the inversion vibration of ammonia; (b) Structure of the ammonia molecule in its two equivalent configurations

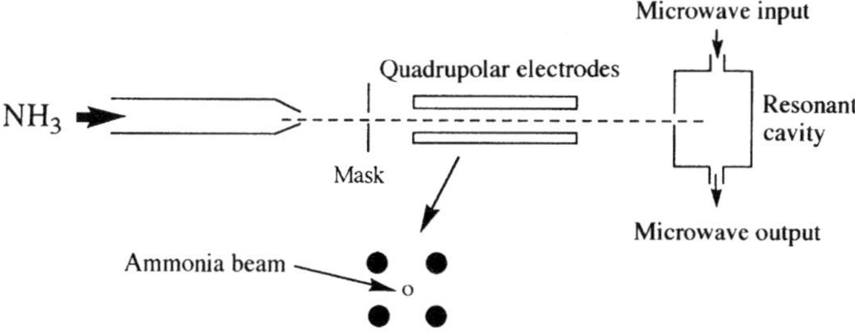

Fig. 19 Experimental setup for the ammonia maser

inversion necessary for maser operation. Within the microwave cavity, which is tuned to resonance with the frequency of the transition, the stimulated emission can be employed to amplify a microwave signal, or to produce oscillation at this frequency.

This description of the ammonia maser is, of course, oversimplified, as its design and construction require considerable care. It is most important to have as many excited molecules as possible entering the cavity per second. Furthermore, the quality factor Q of the cavity should be large and the cavity resonant frequency must be very carefully tuned to that of the inversion line of the ammonia molecule.

The ammonia maser is more suited to its role as a narrow-band oscillator than as an amplifier. The condition for continuous oscillation is that the microwave power produced as stimulated radiation from the molecular beam must be equal to, or greater than, the power lost because of absorption by the cavity. As an oscillator, it produces a stable microwave signal of very low linewidth (approximately 0.01 Hz) if the cavity is carefully designed to avoid temperature, and thus resonant-frequency, fluctuations. A frequency stability of the output signal of one part in 10^{10} over a period of one hour has been achieved. These characteristics make it an ideal frequency standard. The output power obtained is typically of the order of 10^{-10} W.

12.5   CHEMICAL REACTIONS IN CROSSED BEAMS

A particularly ingenious application of molecular beams is in the investigation of chemical reactions. With the use of two crossed beams and a mobile detector, specific information concerning reaction mechanisms can be obtained. A typical experimental arrangement is shown in Fig. 20. In the earliest experiments [13] a beam of potassium atoms was produced by effusion from an oven (A), collimated by slits (S) and velocity selected, as explained above. Another molecular beam of, say, Br_2 molecules produced by a second oven (B) crossed the atomic beam

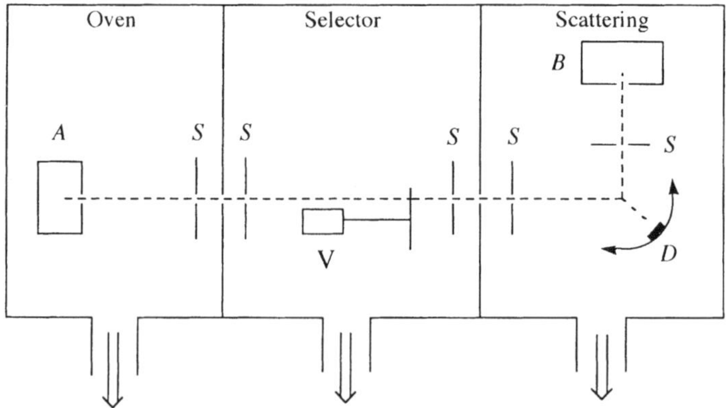

Fig. 20   Crossed-beam experiments for the analysis of chemical reactions (A and B: ovens; S: collimating slits; V: velocity selector; D: movable detector)

(usually at an angle of 90°). A mobile detector (D) was used to analyze the scattered products.

The experiments employing this method were initially limited to reactions of the type

$$A + BC \rightarrow AB + C, \tag{48}$$

where A is potassium, or another alkali metal, and BC is a small molecule, often diatomic. In general, the techniques involved in these experiments are very difficult. The analysis of the scattered species is carried out by measuring the surface ionization of a metallic filament. The results are complicated by the presence of elastically scattered A atoms, as well as the products. In some cases, an electric or inhomogeneous magnetic field is employed to select a given product.

The objective of these reactive-scattering experiments is to determine both the direction and magnitude of the recoil velocity vectors $\mathbf{u}$ that carry the products away from the center of mass. The vector $\mathbf{C}$, which represents the velocity of the center of mass, remains constant and is independent of the result of the collision. The observable distribution of velocity vectors in laboratory coordinates is then given by $\mathbf{v} = \mathbf{u} + \mathbf{C}$, from which the $\mathbf{u}$ vectors can, in principle, be determined.

The results of one of the early investigations of the crossed-beam reaction $K + Br_2 \rightarrow KBr + Br$ are shown in Fig. 21, where the relative intensity is plotted as a function of the scattering angle θ [14]. This angle, which is measured in laboratory coordinates, has a value of approximately 17° at the peak intensity. To interpret the reaction process, the experimental results must be converted into a system of coordinates whose origin is at the center of mass. The vector diagram shown in Fig. 22 provides a useful basis for the analysis of the collision process.

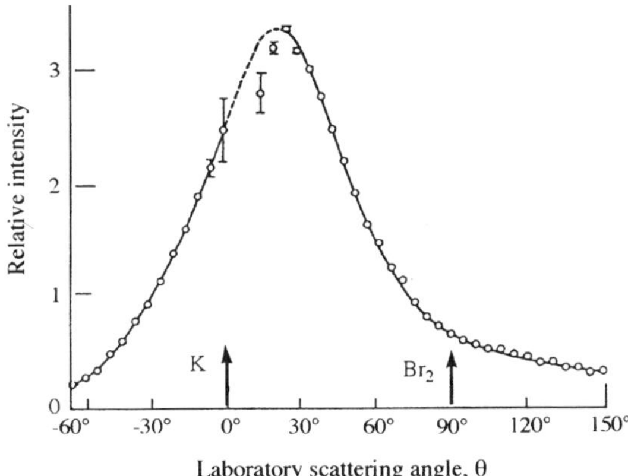

Fig. 21 Results of molecular-beam investigations of the reaction $K + Br_2 \rightarrow KBr + Br$ (from Fig. 5 in Chapter 9 by D. H. Herschbach, *Adv. Chem. Phys.* **10**, 319 (1968), by permission of John Wiley & Sons, New York, NY)

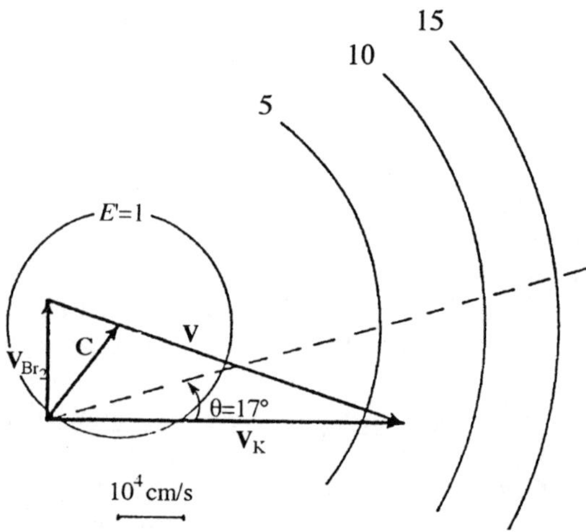

Fig. 22 Interpretation of the $K + Br_2 \rightarrow KBr + Br$ scattering experiments in center-of-mass coordinates (from Fig. 5 in Chapter 9 by D. H. Herschbach, *Adv. Chem. Phys.* **10**, 319 (1968), by permission of John Wiley & Sons, New York, NY)

Here **V** denotes the initial relative velocity vector and **C** is the center-of-mass vector. The arrowhead of **C** divides **V** into two segments whose lengths are inversely proportional to the masses of the reacting molecules. Similarly, the recoil velocity vectors $\mathbf{u} = \mathbf{v} - \mathbf{C}$, are related to the final relative velocity vector **V'** by $\mathbf{u}_{Br} = (m_{HBr}/M)\mathbf{V}'$ and $\mathbf{u}_{HBr} = -(m_{HBr}/M)\mathbf{V}'$, where M is the total mass. Although **V'** can have any direction, its magnitude is determined by the conservation of energy. The total energy available to the reaction products is given by

$$E' + W' = E + W + \Delta D_o, \tag{49}$$

where $E = \frac{1}{2}\mu V^2$ and $E' = \frac{1}{2}\mu' V'^2$ are the initial and final relative translational kinetic energies and μ and μ' are the corresponding reduced masses. In Eq. (49), W and W' denote the internal (vibrational and rotational) energies of Br_2 and HBr, respectively, while ΔD_o is the difference between the dissociation energy of the bond that is formed (H-Br) and that which is broken (Br-Br). For this example, $\Delta D_o \approx 45$ kcal/mol, the maximum in the thermal distribution of initial kinetic energies is at $E \approx 1.2$ kcal/mol, while $W \approx 0.3$ kcal/mol is due to the population of the first excited vibrational state of Br_2. Thus, with the use of Eq. (49), the distribution of recoil vectors $\mathbf{u}_{HBr}$, can be calculated. In Fig. 22 it is represented by a family of spheres centered on the tip of **C**.

In later studies of this reaction, a selector was employed to allow measurements of the velocity distribution of the KBr product [15]. The results of these experiments showed that most of the energy available for this reaction appears in internal excitation of KBr, rather than in translational energy of the products. This result is very important in the analysis of the mechanisms of chemical reactions. It can be summarized as in Fig. 23, where the velocity vector diagram for in-plane scattering is presented. The superimposed curve displaces a

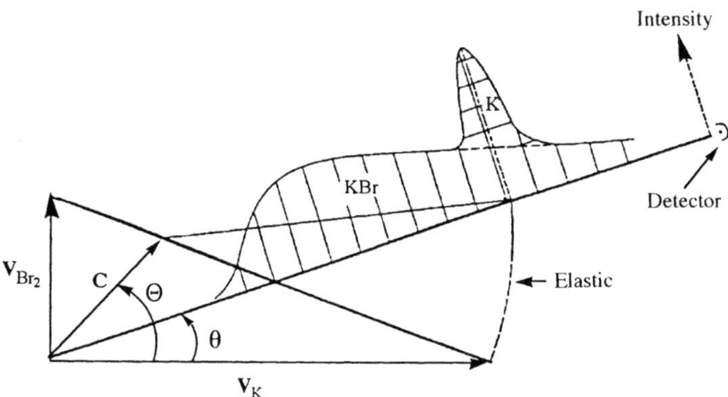

Fig. 23 Analysis of KBr and Br scattering with the use of a velocity selector (reprinted with permission from Fig. 1 in A. E. Grosser and R. B. Bernstein, *J. Chem. Phys.* **43**, 1140 (1965). Copyright (1965) American Institute of Physics, New York, NY)

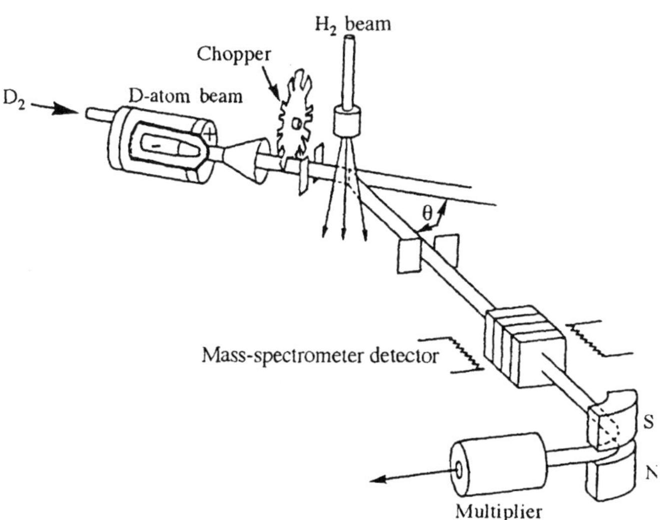

Fig. 24 Experimental arrangement for the investigation of the reaction $D + H_2 \rightarrow HD + H$ (reprinted with permission from Fig. 2 in R. Götting, H. R. Mayne and J. P. Toennies, *J. Chem. Phys.* **85**, 6386 (1986). Copyright (1986) American Institute of Physics, New York, NY)

relatively sharp peak for the K atoms, as opposed to a broad translational energy distribution for KBr.

While experimental work of the type described above has provided the most detailed information concerning chemical reactions, in its initial phases it was limited to reactions involving potassium or other alkali halide atoms. With the refinement of experimental methods, it became possible to study reactions that involve simpler atomic and molecular species. These results were then more amenable to comparison with the results of fundamental theories of reaction mechanisms. For example, *ab initio* calculations of the reaction, say, of deuterium atoms with hydrogen molecules were made, although the corresponding experiments were at the time beyond the capabilities of the available techniques.

Much more recently, very sophisticated experiments have been carried out on the reaction $D + H_2 \rightarrow HD + H$ [16]. The experimental arrangement is shown in Fig. 24. These experiments were made possible by the development of very sensitive time-of-flight (TOF) mass-spectrometer detectors. The results are parti-

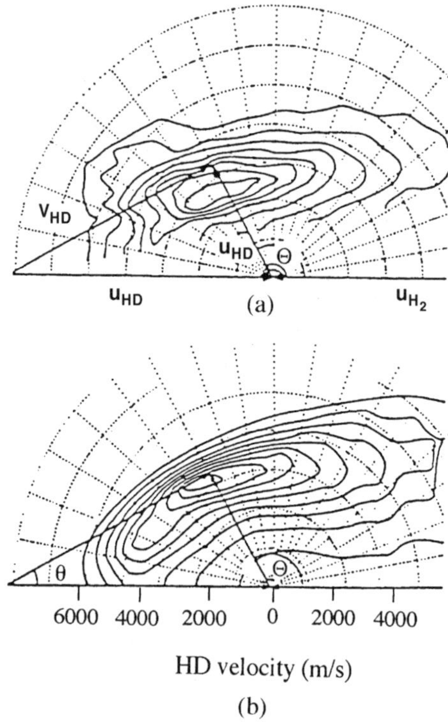

Fig. 25   Contour diagrams for the reaction $D + H_2 \rightarrow HD + H$: (a) Results obtained with the use of the crossed-beam technique; (b) Calculated results (reprinted with permission from Fig. 21 in R. Götting, H. R. Mayne and J. P. Toennies, *J. Chem. Phys.* **85**, 6386 (1986). Copyright (1986) American Institute of Physics, New York, NY)

cularly important as they can be compared with the theoretical studies of these relatively simple species.

To obtain theoretical reaction cross-sections, computer simulations must be carried out over many different trajectories and the results averaged over impact parameters, initial orientations and internal (vibrational and rotational) contributions by both reactants and products. In this way, atomic positions, determined as a function of time, yield specific 'motion pictures' of the formation and decay of the activated complex. The time scale of interest in this case is from about 50 fs to several picoseconds. The results of calculations on the reaction $D + H_2 \rightarrow HD + H$ are summarized in the contour diagram shown in Fig. 25(a) [16]. Here, the flux velocity of the HD product is seen to be directed sideways with respect to the relative velocity. In this figure, Θ is the scattering angle of the product in the center-of-mass system; $\mathbf{u}$ is the velocity in the center-of-mass system and $\mathbf{v}$ is the velocity in the laboratory system.

It is apparent from Fig. 25(a) that the angular distribution of the product formed is quite specific, that is to say, anisotropic in the scattering plane. This result suggests that the activated complex has a very short lifetime, as it does not have enough time to rotate significantly before dissociating into products.

The corresponding experimental results for this reaction, as obtained by the crossed-beam technique, are shown in Fig. 25(b). The agreement of these results with the theoretical predictions is a significant demonstration of the complementarity of experimental and theoretical methods. It is particularly impressive as both the experiments and the calculations are at a high level of sophistication.

Chapter 13

Energy Transfer

The various contributions to the energy of a molecule were considered in Chapter 4. As the molecules in a gas are undergoing collisions, the quantum states of individual molecules are continuously changing. At a given temperature and pressure the system establishes an equilibrium that, however, may be disturbed by an external source. Thus, the passage of a sound wave or the absorption of a light pulse, for example, will upset the equilibrium and result in the relaxation of the system. The characteristic times for such relaxation processes are of considerable practical interest in the analysis of fast chemical reactions, such as combustion, and in hydrodynamic applications. They are also of theoretical importance because they provide information concerning the nature of the intermolecular forces that determine the probability of energy transfer upon collision.

As an example of a perturbation that momentarily disturbs the equilibrium in a gas, consider the passage of a sound wave. It was pointed out by Laplace in 1816 that the propagation of sound in a gas is an adiabatic process. That is, the changes that take place in a gas under the influence of a sound wave are so rapid that insufficient time is allowed for an exchange of energy with the surroundings. It is thus of interest in this chapter to recall the thermodynamics of adiabatic processes and, in particular, the roles of the heat capacities C_V and C_P. These quantities are defined by

$$C_V = \left(\frac{\partial E}{\partial T}\right)_V \tag{50}$$

and

$$C_P = \left(\frac{\partial H}{\partial T}\right)_P, \tag{51}$$

respectively, where $H = E + PV$ is the enthalpy.

The general equation of state of the system can be derived from the fact that $E(P, V, T)$ is constant. Thus, any two of the state variables in parentheses are related, e.g.

$$dE = \left(\frac{\partial E}{\partial T}\right)_V dT + \left(\frac{\partial E}{\partial V}\right)_T dV \tag{52}$$

and

$$dE = \left(\frac{\partial E}{\partial T}\right)_P dT + \left(\frac{\partial E}{\partial P}\right)_T dP. \tag{53}$$

Similarly, since the volume is a function of temperature and pressure,

$$dV = \left(\frac{\partial V}{\partial T}\right)_P dT + \left(\frac{\partial V}{\partial P}\right)_T dP. \tag{54}$$

Substitution of Eq. (54) into Eq. (52) yields

$$dE = \left(\frac{\partial E}{\partial T}\right)_V dT + \left(\frac{\partial E}{\partial V}\right)_T \left[\left(\frac{\partial V}{\partial T}\right)_P dT + \left(\frac{\partial V}{\partial P}\right)_T dP\right] \tag{55}$$

and thus,

$$\left(\frac{\partial E}{\partial T}\right)_P = \left(\frac{\partial E}{\partial T}\right)_V + \left(\frac{\partial E}{\partial V}\right)_T \left(\frac{\partial V}{\partial T}\right)_P. \tag{56}$$

From the definitions given above

$$C_P - C_V = \left(\frac{\partial H}{\partial T}\right)_P - \left(\frac{\partial E}{\partial T}\right)_V \tag{57}$$

$$= \left(\frac{\partial E}{\partial T}\right)_P + P\left(\frac{\partial V}{\partial T}\right)_P - \left(\frac{\partial E}{\partial T}\right)_V \tag{58}$$

and

$$C_P - C_V = \left(\frac{\partial V}{\partial T}\right)_P \left[P + \left(\frac{\partial E}{\partial V}\right)_T\right]. \tag{59}$$

The significance of the two terms in brackets in Eq. (59) is particularly important. The first is the pressure P, which is specifically the external pressure applied on the system. The second represents the change in internal energy with volume at constant temperature. It is the result of intermolecular forces, as described in Chapter 6. Clearly, for an ideal gas, $(\partial E/\partial V)_T = 0$, a relation that can be taken as its fundamental definition. Furthermore, as $PV = n_0 RT$ in this case [Eq. (I.16)],

$$P\left(\frac{\partial V}{\partial T}\right)_P = n_0 R \tag{60}$$

and

$$C_P - C_V = n_0 R \tag{61}$$

or, in terms of molar quantities,

$$\tilde{C}_P - \tilde{C}_V = R. \tag{62}$$

The heat-capacity ratio $\gamma = \tilde{C}_P / \tilde{C}_V$ is generally used as a basic parameter. Thus, for an ideal gas

$$\gamma = \frac{(R + \tilde{C}_V)}{\tilde{C}_V} = 1 + \left(\frac{R}{\tilde{C}_V} \right). \tag{63}$$

From the temperature dependence of $\tilde{C}_V$, as illustrated in Fig. I.11, the variation of γ with temperature can be determined. Some experimental values for some simple gases are given in Table 2. The classical limit to γ can be calculated from the expressions given in Table I.2.

13.1   VELOCITY OF SOUND

It should be recalled that for an adiabatic process such as the propagation of sound, $dE = -P\,dV$ and for an ideal gas, Eq. (52) becomes

$$P\,d\tilde{V} + \tilde{C}_V\,dT = 0, \tag{64}$$

or

$$\frac{R\,dV}{V} + \frac{\tilde{C}_V\,dT}{T} = 0. \tag{65}$$

Integration of Eq. (65) for an adiabatic process between $T_1 V_1$ and $T_2 V_2$, assuming that $\tilde{C}_V$ is constant over a limited temperature range, yields

$$R\ln\frac{V_2}{V_1} + \tilde{C}_V \ln\frac{T_2}{T_1} = 0. \tag{66}$$

However, it should be noted that $\tilde{C}_V$ is not, in general, independent of temperature (see Fig. I.11). With the aid of Eq. (63) this result can be written in the form

$$\left(\frac{V_2}{V_1} \right)^{\gamma-1} = \frac{T_1}{T_2}, \tag{67}$$

which leads to the basic relation, $PV^\gamma = \text{constant}$.

Table 2   Ratios of heat capacities ($\gamma = C_P / C_V$)

Gas	Temperature (K)				γ, Classical limit
	150	300	500	900	
O_2	1.40	1.39	1.37	1.32	1.29
H_2	1.45	1.40	1.39	1.38	1.29
H_2O	–	1.33	1.31	1.26	1.17
CO_2	–	1.29	1.23	1.19	1.15
Ar	1.67	1.67	1.67	1.67	1.67

A plane sound wave in the z direction in a gas of density ρ can be represented by

$$\rho = \rho_0(1 + \zeta), \tag{68}$$

where

$$\zeta = \zeta_0\, e^{-2\pi i(\alpha z - \nu t)}. \tag{69}$$

In Eq. (69), ζ_0 is the amplitude of the wave, $\alpha = 1/\lambda$ its wave number and ν its frequency. If the wave, as shown in Fig. 26, is considered to be stationary, then its propagation in the gas corresponds to a flow of gas with a speed equal to c. If the process takes place in a hypothetical tube of unit cross-section, then the mass traversing section **a** per second is equal to $\rho_a c_a$. Thus, as the mass is conserved, $\rho_a c_a = \rho_b c_b$. Then, the rate of change of momentum per second at **a** is $(\rho_a c_a)c_a$, with an analogous expression for the momentum at **b**. Since a unit cross-section has been assumed, according to Newton's second law, the difference between these quantities is equal to the pressure difference between the sections **a** and **b**. The result is then given by

$$\rho_b c_b^2 - \rho_a c_a^2 = P_a - P_b, \tag{70}$$

which can be written as

$$P_a - P_b = \rho_a^2 c_a^2\left(\frac{1}{\rho_b} - \frac{1}{\rho_a}\right) = \frac{Mc_a^2}{\tilde{V}_a^2}(\tilde{V}_b - \tilde{V}_a). \tag{71}$$

Then,

$$c_a^2 = \frac{\tilde{V}_a^2}{M}\left(\frac{P_a - P_b}{\tilde{V}_b - \tilde{V}_a}\right) \tag{72}$$

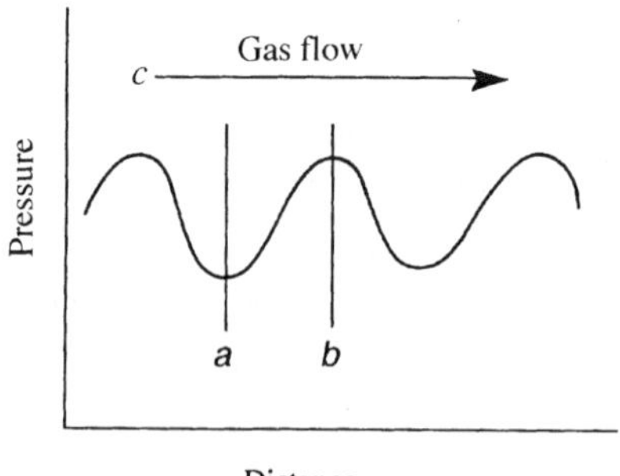

Fig. 26 Stationary planar sound wave in gas

and, for a sound wave of infinitesimal amplitude, c_a can be replaced by c, yielding,

$$c^2 = \frac{\tilde{V}^2}{M}\left[-\left(\frac{\partial P}{\partial \tilde{V}}\right)_{ad}\right]. \tag{73}$$

The subscript 'ad' in Eq. (73) indicates that the process is adiabatic, as it is too rapid to permit the exchange of heat between the regions of compression and rarefaction. Thus, temperature equilibrium is not achieved during the passage of the sound wave.

For an adiabatic process in an ideal gas, $P\tilde{V}^\gamma$ is constant; then,

$$\left(\frac{\partial P}{\partial \tilde{V}}\right)_{ad} = -\frac{\gamma P}{\tilde{V}} \tag{74}$$

and the sound velocity is given by

$$c = \sqrt{\frac{\gamma RT}{M}}. \tag{75}$$

For an ideal gas, the heat-capacity ratio γ is given by Eq. (63).

In the analysis of sound propagation presented above it was assumed that the sound wave was not absorbed by the gas. However, if so-called dissipative processes are present, then sound absorption will occur that depends on the various transport coefficients, possible chemical reactions and relaxation phenomena. In molecules with internal degrees of freedom, the exchange of energy during collisions—the energy relaxation—leads to sound dispersion, as well as absorption.

A one-dimensional sound wave in a gas was described by Eqs. (68) and (69). The quantity α was defined as the wavenumber. However, in the presence of absorption, the wavenumber becomes a complex quantity that can be written as $\alpha + i\kappa$. The absorption coefficient κ is non-zero in the presence of dissipative processes such as relaxation.

The analysis of the relaxation of an internal degree of freedom in a molecule can be made by analogy with a chemical reaction. The gas is assumed to be composed of two types of molecule: unexcited molecules A, and molecules A^* that possess an additional energy ε^* in an internal degree of freedom. The transfer of energy can then be represented by

$$M + A^* \rightarrow M + A, \tag{76}$$

where M may be a molecule of either A or A^*. This first-order process is described by

$$-\frac{d\varepsilon^*}{dt} = k(\varepsilon^* - \varepsilon_{tr}), \tag{77}$$

where ε_{tr} is the translational energy and k is the rate constant. It is convenient to define the relaxation time by

$$\tau = 1/k = 1/(Zp) \tag{78}$$

(see Chapter 10). Here, Z is the number of collisions per second and p is the probability of energy transfer as a result of a given collision.

The quantity $k = 1/\tau$ defined above can be considered to be a frequency that is characteristic of the energy-transfer process. At sound frequencies above this value the compressions and rarefactions take place so rapidly that the internal degree of freedom being considered cannot contribute to C_V and, hence, to the heat-capacity ratio. At low frequencies, however, the rate of energy exchange is rapid enough with respect to the sound frequency to allow equilibrium to be achieved at each period of the sound wave. The result is the so-called dispersion of sound, e.g. the frequency dependence of the sound velocity, as illustrated in Fig. 27(a). The inflection point on this curve is given approximately by

$$v_\tau = \frac{1}{2\pi\tau} \frac{C_V(\text{low})}{C_V(\text{high})}, \tag{79}$$

where $C_V(\text{high})$ and $C_V(\text{low})$ refer respectively to the heat capacities at higher and lower sound frequencies. Analogous results can be obtained for the sound absorption. As shown in Fig. 27(b), the maximum in the absorption occurs at a frequency that is close to that given by Eq. (79).

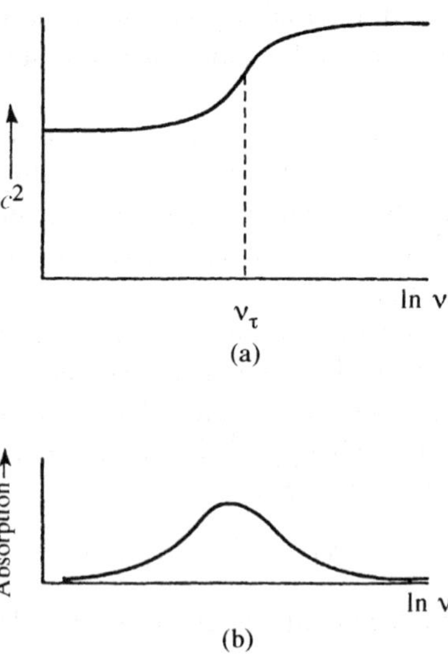

Fig. 27 Dependence of the velocity (a) and the absorption coefficient (b) on the frequency of a sound wave

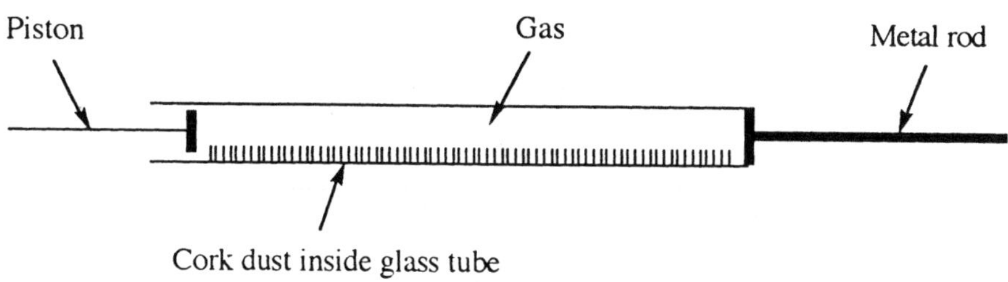

Fig. 28 Kundt's method for the measurement of the sound velocity in a gas

In the general physics laboratory the sound velocity in a gas is usually measured by the method introduced by Kundt. A glass tube of a few centimeters in diameter and a meter or so long is filled with a gas (see Fig. 28). One end of the tube is closed, while the other is fitted with a piston, so that the total interior length of the tube can be adjusted. The metallic shaft attached to the piston is acoustically activated, either manually (a chamois dusted with rosin is tradition- ally employed), or with an audio oscillator, to generate a sound wave in the tube. A small quantity of cork dust is introduced into the tube before filling with the gas. The cork dust serves to identify the positions of the maxima and minima in the sound wave. This powder collects in small heaps at the nodes, where the gas is not moving, and produces clear spaces at the antinodes, where the moving gas molecules sweep away the powder. The wavelength of the sound wave in the gas is then measured from the distance between the nodes. More sophisticated methods that employ electronic techniques are used in research experiments to measure sound absorption, as well as dispersion, as a function of temperature.

These experiments provide a direct method for the measurement of relaxation times; in particular, the vibrational relaxation of simple molecules. In principle, the data yield information concerning the intermolecular forces that determine the interaction of molecules in collision. However, as the theoretical background to this problem is relatively complicated, the quantitative interpretation of such experimental results is not obvious.

13.2 SHOCK WAVES

In Section 13.1, sound waves were considered to be of infinitesimal amplitude. The sound velocity was treated as a thermodynamic quantity, as given by Eq. (75). However, in a real fluid, as the intensity of the wave becomes greater, the sound velocity increases with pressure along an adiabat. The fact that the speed of sound is greater in the high-pressure region of a sound wave explains, at least in part, the formation of shock waves.

Consider the sinusoidal, one-dimensional wave, as depicted in Fig. 29(a). If the density is significantly higher in the high-pressure region, then the crest of

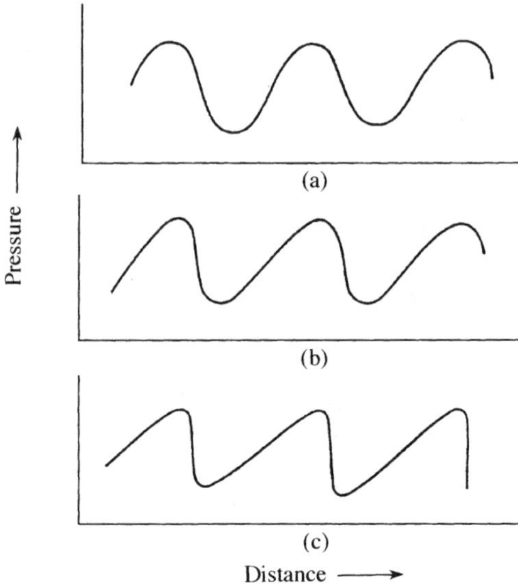

Fig. 29   (a) Weak sound wave in a gas (sinusoidal); (b), (c) Development of a shock wave in a strong sound wave

the wave will travel faster than the trough. After some time, the pressure profile of the wave will begin to take the form shown in Fig. 29(b). Finally, the sharp discontinuity at the front of the resulting wave (Fig. 29(c)) describes a shock wave in which the density change occurs within a period of time that corresponds to relatively few molecular collisions.

The classical analog of the development of shock waves in gases is provided by the motion of waves in the ocean as they approach shallow water. The velocity of a wave in water is proportional to its depth. Thus, the top of a wave travels at a higher speed than the water at a lower level. Then, the leading edge should, in principle, become vertical, although before this stage is reached a breaker is formed and the energy is dissipated, often quite violently, on the beach.

From the point of view of chemists and physicists the shock-wave technique provides a method of increasing the temperature of a gas to very high temperatures within a period of time of nanoseconds, or less. The gas is thus heated rapidly and homogeneously to very high temperatures. The increase in temperature across the shock wave provides a method to study the various mechanisms of energy exchange between the degrees of freedom of the molecules present in the system and their approach to equilibrium. With the use of sufficiently strong shock waves, processes such as dissociation, electronic excitation and ionization, as well as rotation and vibration, can be investigated..The method has also been applied to the study of chemical reactions involving relatively simple molecules [17].

In engineering applications, aerodynamic studies of shock waves are usually made with a model, such as an aircraft, mounted in a wind tunnel. Optical methods, such as schlieren photography,* are employed to observe the density profiles of stationary shock waves. However, in fundamental laboratory experiments, shock waves are usually generated with the use of a shock tube, as shown in Fig. 30. The tube shown is divided into two sections by a plastic or thin metallic diaphragm. The gas to be studied is introduced at a relatively low pressure into one section, while the other is filled with the same, or another (the driver) gas. The pressure of the driver gas is then increased until the diaphragm breaks, or it is punctured upon arrival at the desired pressure by a needle introduced into the system.

After the rupture of the diaphragm, the shock wave travels along the tube at a supersonic velocity. This velocity depends on the relative pressures in the two sections of the shock tube before the initial event, the rupture of the diaphragm at $t = 0$. The relations between the initial pressure ratio and the properties of the shock wave are relatively complicated. However, they are derived from the basic principles of the conservation of mass, momentum and energy of the interacting

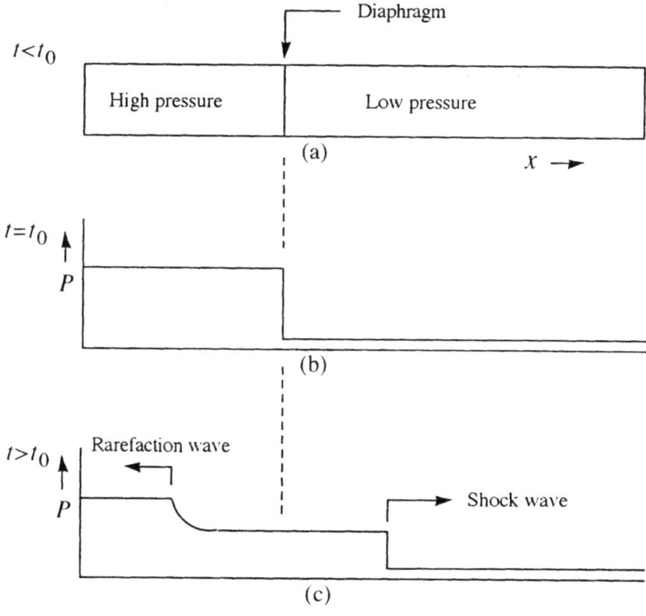

Fig. 30   The shock tube (a) and the pressure profiles before (b) and after (c) rupture of the diaphragm

---

*The schlieren effect results from refractive-index changes associated with density gradients in a fluid.

molecules. The first two of these relations were written in Section 13.1. With the addition of the condition for the conservation of energy, the three basic relations can be expressed in the form

(i) $\rho_a c_a = \rho_b c_b$, for the conservation of mass,   (80)

(ii) $\rho_a c_a^2 + P_a = \rho_b c_b^2 + P_b$, for the conservation of momentum [Eq. (70)]
$$\tag{81}$$

and

(iii) $\tfrac{1}{2}c_a^2 + h_a^2 = \tfrac{1}{2}c_b^2 + h_b.$ (82)

In Eq. (82) the enthalpy per gram is given by $h = \tilde{H}/M$, where M is the molecular weight. Since $H = E + PV$, Eq. (82) can also be written as

$$\tfrac{1}{2}c_a^2 + \varepsilon_a + \frac{P_a}{\rho_a} = \tfrac{1}{2}c_b^2 + \varepsilon_b + \frac{P_b}{\rho_b}, \tag{83}$$

with $\varepsilon = \tilde{E}/M$. These conditions, which yield the fundamental relations between the thermodynamic quantities involved, are often ascribed to Hugoniot. The value of c_a is the velocity of the cold gas into the shock or the velocity of the propagation of the shock wave into stationary gas on the low-pressure side.

The substitution of the expression for c_b from Eq. (80) into Eqs (81) and (83), after a certain amount of algebra, will lead to the Hugoniot equation in the form

$$\varepsilon_b - \varepsilon_a = \tfrac{1}{2}(P_b + P_a)\left(\frac{1}{\rho_a} - \frac{1}{\rho_b}\right). \tag{84}$$

This relation can be used to determine the temperature rise across a normal shock wave if the relations between internal energy and temperature are known and no chemical reactions are involved. Usually, the density and pressure on the low-pressure side of the shock wave are specified. In addition, a measure is made of the strength of the shock wave—the pressure on the high side, or, more often, an experimental determination of the shock velocity.

For an ideal gas, the ratio of heat capacities can be written in the form $\gamma = 1 + R/\tilde{C}_V$, as given by Eq. (63). Then, the internal energy per unit mass at each side of the shock wave can be expressed as

$$\varepsilon = \frac{1}{M}\left(\frac{\partial \tilde{E}}{\partial T}\right)_V, \quad T = \frac{P}{(\gamma - 1)\rho}. \tag{85}$$

Substitution of Eq. (85) into Eq. (84) leads to the relation

$$\frac{\rho_b}{\rho_a} = \frac{(\gamma - 1)P_a + (\gamma + 1)P_b}{(\gamma + 1)P_a + (\gamma - 1)P_b}. \tag{86}$$

For strong shock waves, $P_b \gg P_a$ and the density ratio given by Eq. (86) approaches $(\gamma + 1)/(\gamma - 1)$. As an example, for air ($\gamma = 1.4$) this limiting value is equal to six. For an ideal gas, the temperature ratio across the shock front is given by

$$\frac{T_b}{T_a} = \left(\frac{P_b}{P_a}\right)\left(\frac{\rho_a}{\rho_b}\right). \tag{87}$$

Thus, as the density ratio is limited, the temperature produced by the shock wave can become extremely high.

Equation (81) can be rearranged in the form

$$\frac{P_b - P_a}{\rho_b - \rho_a} = \gamma\frac{P_b + P_a}{\rho_b + \rho_a}, \tag{88}$$

where it is known as the Rankine–Hugoniot equation. It should be noted that for weak shock waves, $P_b \to P_a$ and Eq. (88) becomes

$$\left(\frac{\partial P}{\partial \rho}\right) = \gamma\frac{P}{\rho}. \tag{89}$$

Equation (89) applies to an adiabatic process in an ideal gas, in which the entropy is conserved.

In practical applications, it is useful to have an expression for the velocity of the shock wave with respect to the still gas on the low-frequency side. From Eqs (80) and (81),

$$c_a = \sqrt{\frac{\rho_b}{\rho_a}\frac{P_b - P_a}{\rho_b - \rho_a}} \approx \sqrt{\frac{P_b - P_a}{\rho_a}}, \tag{90}$$

where the approximation is appropriate for strong shock waves. The ratio of the value calculated from Eq. (90) to the velocity of sound in the still gas is the Mach number, the usual measure of the relative shock-wave velocity.

A simple analysis of the contributions to the heat capacity C_V of the internal degrees of freedom of rotation and vibration of a molecule was presented in Chapter 4. If, at a given time, the various degrees are not in equilibrium, then it is necessary to redefine the concept of temperature. The temperature, as specified by the translational kinetic energy of the system, is then different from the effective temperatures associated with the internal degrees of freedom of the molecule.

With the passage of a shock wave, the molecules immediately obtain the translational energy of the driver gas. However, although this effect of the shock front is virtually instantaneous, it has been shown by optical reflection that between five and ten collisions are normally necessary to establish equilibrium with the rotational degrees of freedom of, say, a diatomic molecule [18]. Thus, the rotational temperature becomes equal to that of the translational degrees of freedom within a few molecular collisions. However, the vibrational degree of freedom in this case requires relatively longer times to reach equilibrium. The general form of the pressure profile of a shock wave in this case is illustrated in Fig. 31.

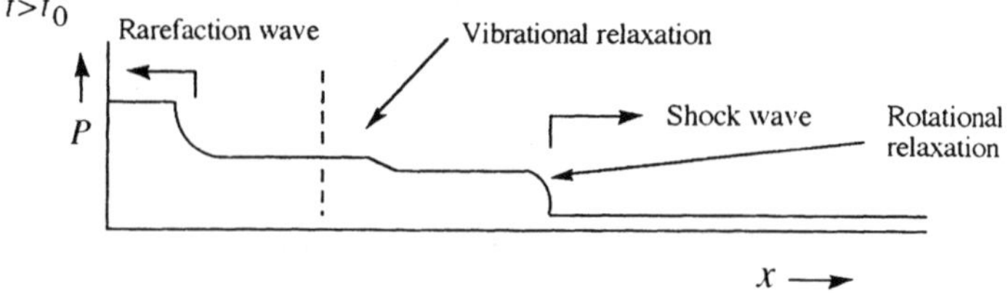

Fig. 31   Pressure profile in a shock wave showing the effects of rotational and vibrational relaxation

Experimental measurements of the shock-wave velocity, traditionally expressed by the Mach number, are made to determine the pressure ratio developed by the wave. They are usually combined with additional observations, such as the pressure profile (as measured by very sensitive and rapid detectors), absorption or emission spectra, optical reflection, etc. It should be apparent that there is a distinct similarity between shock-wave investigations and those involving flash photolysis, as described in Chapter 10. However, the latter method can provide neither the rapidity of heating nor the uniformity of temperature distribution obtained with a shock wave.

The analysis presented here of the properties of a simple shock wave can also be applied to gaseous combustion. Consider a mixture of combustible gases such as hydrogen and oxygen. It can be ignited, for example, by an electric spark at one end of the shock tube. In this case, the reaction is highly exothermic, a detonation will be produced and a shock front will be propagated at a supersonic velocity. For the analysis of this process, the Hugoniot relation [Eq. (83)] can be modified by the addition of the (negative) heat of the reaction on the left-hand side. In the case of a reaction that is not sufficiently exothermic, the result is deflagration, which yields a wave propagating at a subsonic velocity.

13.3   THE OPTICAL–ACOUSTICAL EFFECT

In 1880, Alexander Graham Bell, the acknowledged inventor of the telephone, published the results of his experiments on the interaction of light and sound [19]. He devised an apparatus that consisted of a beam of sunlight interrupted by a rotating sector or 'chopper', which served to modulate the incident light at an audio frequency. This beam of light was focused on a sample contained in a suitable cell. With the aid of a stethoscope, Bell listened to the sound produced in the cell at the frequency of the light modulation. He called this apparatus the 'spectrophone'. This experiment with the spectrophone was repeated on a large number of substances in the cell and, with the aid of a prism, different wavelengths of the incident light were chosen and the relative intensities of the sound evaluated. These experiments represent the first application of what is now called photoacoustic spectroscopy.

Within the context of this chapter, Bell's experiments demonstrate directly the exchange of energy within a molecular system. As shown in Fig. 32, the modulated light beam falls on, say, a gaseous molecule and is absorbed. If the light is of a suitable frequency; for example, in the infrared spectral region, the result is to excite the molecule to a higher vibrational level (see Fig. 33). Such a vibrationally excited molecule will then undergo collisions with its neighbors. If a certain number of these collisions are inelastic, the excess energy of vibration will be transferred into translational energy, resulting in an instantaneous increase in temperature. In a system at constant volume, a pressure increase will be created. Hence, by chopping the incident light, a pressure wave will be produced

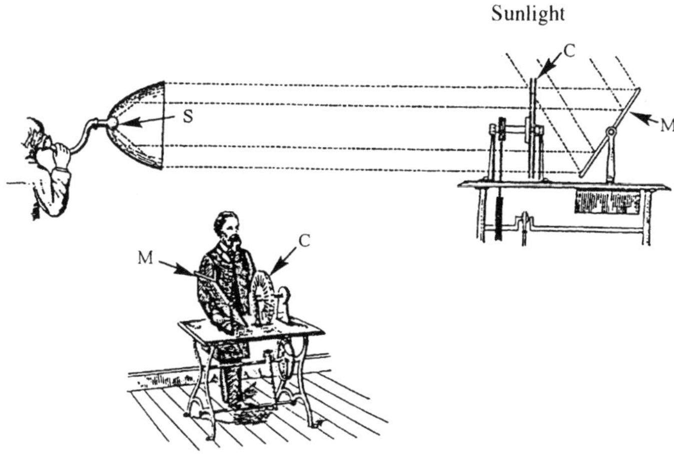

Fig. 32 The photoacoustic experiment of Alexander Graham Bell (1880). *M*: plane mirror; *C*: chopper; *S*: sample in bulb (reproduced from Fig. 3 in A. G. Bell, *Phil. Mag.* **11**, 510 (1881), Taylor and Francis, Ltd, London)

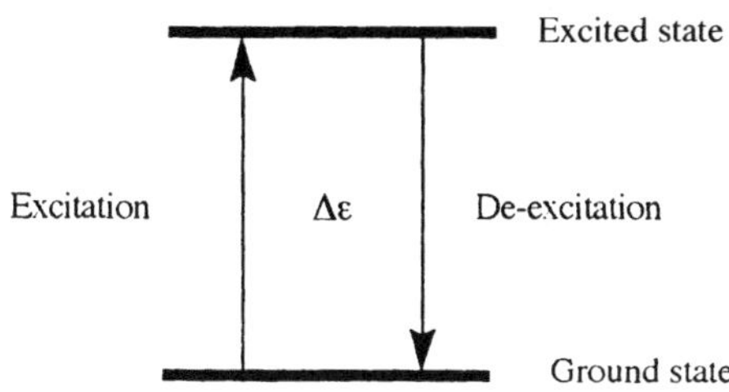

Fig. 33 Excitation and de-excitation processes in a two-level system

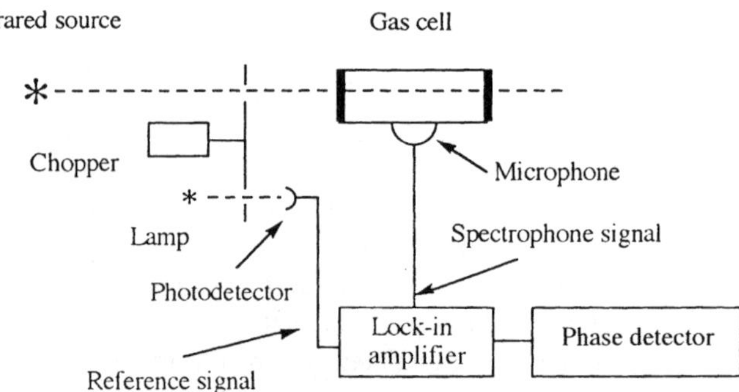

Fig. 34   Schematic diagram of the spectrophone

as a sound wave at the frequency of the light modulation. It was this signal that Bell heard in his stethoscope.

The spectrophone was forgotten until the 1940s, when Soviet scientists attempted to determine the lifetime of the excited vibrational state of CO_2 by measuring the time delay, or phase shift, between the arrival of the light and the detection of the sound [20]. The basic spectrophone system used in this application is illustrated in Fig. 34. These experiments were repeated and developed in the United States [21], and later in England [22], in an effort to measure directly the relaxation time of the vibrational energy of a molecule. Most of the early experiments were failures, as most experiments are! In general, the researchers forgot that the loss of excess energy by a molecule as a result of an inelastic collision is in competition with its loss by radiation. Thus, only at relatively high pressures, where collision mechanisms dominate, will the phase shift in the spectrophone be determined by the relaxation time of the vibrational energy.

In spite of the lack of success of the spectrophone for relaxation-time measurements, the optical–acoustical effect enjoyed a renaissance with the development of Fourier-transform infrared spectroscopy. In what is now known as photoacoustic spectroscopy, the analytical use of this method has become important, particularly in industrial applications. For example, it has been shown to be useful in the investigations of the surfaces of bulk materials, catalytic surfaces and adsorbed species.

Chapter 14

Broadening of Spectral Lines

According to quantum theory, atomic and molecular energies are quantized, a specific state or states being associated with a given value of the energy. Some examples for a simple molecule were summarized in Chapter 4. The interaction of electromagnetic radiation—light, in the general sense—with matter involves a change in state of the atoms or molecules of which it is composed. Hence, according to Planck, the energy associated with such a change is related to the light frequency by $\varepsilon_{nm} = h\nu_{nm}$, where h is Planck's constant and ν_{nm} is the frequency associated with a transition between energy levels m and n. Here, ε_{nm}, which is the difference in energy between the states m and n of the system, would be expected to have a specific value. The spectrum of the system, as observed in absorption or emission, for example, should then exhibit one or more 'lines' of zero width at frequencies ν_{nm}.[*] However, even in the case of relatively low-pressure gaseous samples, several physical phenomena are involved that broaden a given spectral line.

14.1 UNCERTAINTY BROADENING

The most fundamental source of line broadening is the result of Heisenberg's uncertainty principle. Although this principle is usually described as an uncertainty in the simultaneous determination of the position and the momentum of a particle, in the present context it is expressed by the analogous uncertainty between the time of a given event and the energy involved. Thus, with ε_{nm} the energy of a given transition, $\delta\varepsilon_{nm}\delta t \approx h/2\pi$, where the symbol δ represents the uncertainty in the physical quantity indicated. For an uncertainty in the time of the transition there is then a related uncertainty in the energy. The result is a

---

[*] The term 'line' originated early in the development of spectroscopic techniques, when photographic plates were used to record spectra. The distinct features observed on the plates, the so-called lines, were, in fact, the images of the optical slit employed in the monochromator.

broadening of the spectral feature associated with the transition. The theoretical analysis of this problem leads to the expression for the line shape in the form

$$\mathscr{S}(\nu) = \frac{a}{(\nu - \nu_{nm})^2 + b^2}, \tag{91}$$

where a and b are constants. This function describes what is known as a Lorentzian band shape. The maximum occurs at $\nu = \nu_{nm}$, where it has the value a/b^2. At half of this maximum value, $\nu = \nu_{nm} \pm \Delta\nu$, where $\Delta\nu$ is half the width of the line at half maximum (HWHM). It should be noted that, in practice, the quantity FWHM is often employed. It is the full width at half maximum, which is equal to $2\Delta\nu$, as defined here. From Eq. (91) it can be easily shown that $\Delta\nu = b$, the HWHM of the spectral line.

The broadening of spectral lines as a result of the uncertainty principle, which is often referred to as natural line broadening, is, in general, quite small. In the microwave region, for example, this effect results in a broadening of the order of 10^{-4} Hz for rotational transitions, while at optical frequencies it becomes greater. Nevertheless, uncertainty broadening of spectral lines is usually much less important than the results of other effects, as discussed in what follows.

14.2   THE DOPPLER EFFECT

The Doppler effect is usually described as an acoustical phenomenon. Anyone who drives at a reasonable speed along a highway knows that if a fire engine approaches from behind, the frequency of its siren will decrease at the moment when it passes. The analogous problem arises with the consideration of the interaction of a molecule with a beam of electromagnetic radiation. The vibrational frequency of the molecule at the moment of its interaction will depend on the velocity of translational movement of the molecule with respect to the direction of the light beam. This problem can be analyzed in a simple way with the use of the one-dimensional distribution of molecular speeds, as developed in Chapter 3.

The effective vibrational frequency of a molecule moving perpendicular to the direction of propagation of the light beam will be just that of a stationary molecule, ν_0. However, molecules moving parallel to the light at a speed u will have observable frequencies given by

$$\nu = \nu_0\left(1 \pm \frac{u}{c}\right), \tag{92}$$

where the sign is determined by the direction of displacement with respect to that of the light beam. According to Eq. (I.47) the distribution of molecular speeds of a molecule of mass m in, say, the x direction is proportional to $\exp(-mu^2/2kT)$, where $\dot{x}$ has been identified with u in this one-dimensional application. Substitution of u from Eq. (92) leads to a frequency distribution of the form

$$\mathscr{I}(\nu) = \mathscr{B} \exp\left[-\frac{mc^2(\nu - \nu_0)^2}{\nu_0^2 2kT}\right], \tag{93}$$

where $\mathscr{B}$ is a constant. Equation (93) represents a Gaussian line shape with a maximum equal to $\mathscr{B}$ at $\nu = \nu_0$ and a width (HWHM) given by

$$\Delta\nu = \frac{\nu_0}{c}\sqrt{\frac{2kT \ln 2}{m}}. \tag{94}$$

Under normal experimental conditions, Eq. (93) corresponds to a line that is much larger than the natural width due to the uncertainty principle.

14.3 PRESSURE BROADENING

In the analyses of spectral line shapes, as outlined above, it was assumed that each molecule in the system is isolated. All molecular interactions; namely, collisions, were neglected. In a gaseous sample, even at relatively low pressures, the effect of intermolecular forces must be considered. The origin of these interactions was considered in Chapter 6. Their role in the trajectories of molecular collisions was briefly summarized in Chapter 8, albeit with simplifications. If, as a first approximation, the angular dependencies of the interaction forces can be neglected, then a typical binary interaction can be represented by Fig. I.21, where the coordinate system is chosen so that the molecule of interest, which is at rest, suffers a collision with another molecule. In pressure-broadening experiments, the molecule of interest may be a diatomic, such as HCl or CO, while a 'foreign' molecule, say, Ar, is introduced to perturb the spectrum of the molecule of interest.

Consider a simple example in which the molecule of interest is diatomic with a characteristic vibrational frequency ν_0, as indicated in Chapter 4. The effect of a 'collision' with a foreign atom or molecule will be to modify, or even effectively stop, the vibration of the diatomic molecule during the period of binary interaction. Thus, the time during which the molecule enjoys free vibration at frequency ν_0 is limited to the period τ between successive collisions with the foreign perturber. This situation is represented schematically in Fig. 35. The resulting spectrum, even if broadening occurs because of the uncertainty principle and the Doppler effect, will be more severely modified by successive molecular collisions.

If the intermolecular perturbation can be described as a function of time, then the resulting spectral profile, or band shape, is the Fourier transform of this time-dependent function. To explain in simple terms the physical significance of the Fourier transform, it is interesting to consider the following 'Gedankenexperiment'. This is the sort of experiment that is carried out with the application of thoughts (Dänken), rather than hands (manipulations).

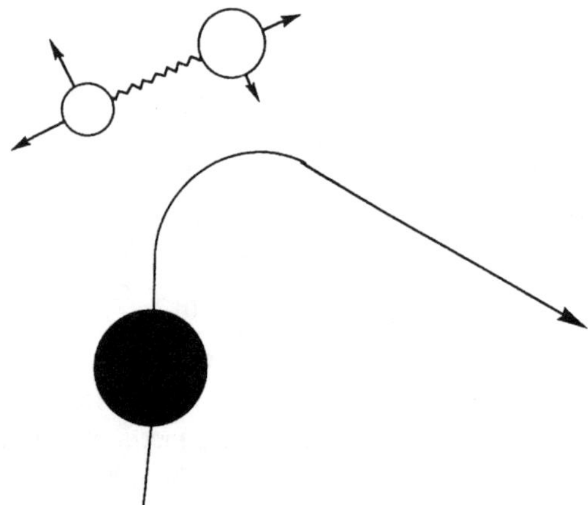

Fig. 35   Collision between a monatomic and a diatomic molecule

A young research student has just received an oscillator that is claimed by the manufacturer to generate a frequency of 10 MHz. He plugs in the 'black box' when he arrives in the laboratory. He then goes out for his morning coffee break. On his return to the laboratory he connects the output of the black box to his oscilloscope and observes a beautiful sine wave. Knowing the time base of his oscilloscope, he immediately verifies that the signal is, indeed, at a frequency of 10 MHz.

One morning the student comes in late and turns on his oscilloscope. He connects the black box and immediately looks at the screen. If he is sufficiently awake, he may notice that the oscillations shown on the screen are initially somewhat perturbed. It may not be his fault, for, in fact, the frequency of the oscillator depends on the length of time that the signal has been produced.

To illustrate this effect, consider the case in which the oscillator is turned on at a time $t = 0$ and allowed to oscillate until a time $t = \tau$, when it is turned off. The envelope of the oscillations then corresponds to what is often referred to as a boxcar (see Fig. 35(a)). Its Fourier transform corresponds to a function of the form $(\sin v)/v$, which is usually called sinc v. This function is plotted if Fig. 36(b), where it is shown that the width of the frequency distribution varies inversely with τ. The energy spectrum, as would be observed with a spectrometer, is the square of this distribution.*

---

* Students who have had the occasion to use a Fourier-transform infrared spectrometer may have noticed that an absorbtion band sometimes displays weak bands on either side. These features, which have been described as 'Gibbs' ears' (after J. Willard Gibbs), are artifacts arising from the finite distance of displacement of the moving mirror in the interferometer. They are usually removed by the suitable choice of an apodizing function.

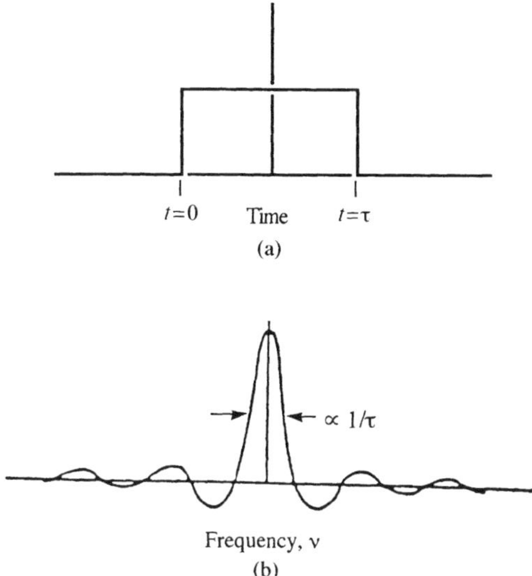

Fig. 36   (a) Boxcar function of duration τ; (b) corresponding frequency spectrum

Returning now to the problem of the collision of the vibrating diatomic molecule with a foreign atom, it will be assumed for simplicity that a spherically symmetric potential function can be used to represent the intermolecular interaction. The trajectory of a bimolecular collision might then be represented by Fig. 35. As the interaction gradually increases, the perturbation of the vibrational motion of the molecule of interest increases accordingly until a maximum is reached. The maximum corresponds to the distance of closest approach of the perturbing molecule. As time goes on, the perturbation decreases. The general form of this interaction is at least approximately symmetric, yielding by Fourier transformation a symmetrical spectral line. This profile, which becomes very wide at high pressures, is often referred to as a band. Its shape can often be approximated by a combination of Gaussian and Lorentzian functions (the Voigt function). Modern infrared spectrometers are usually equipped with software that permits curve fitting to be made to an observed absorption band with the aid of this function.

In principle, the measurement of pressure-broadening effects on band shapes provides information concerning the intermolecular forces that govern the collisions. For spherical molecules, the number of collisions per second that the molecule of interest makes with all others is given by Eq. (I.62). The time between successive collisions is then given by

$$\tau = \frac{1}{Z_a} = (\sqrt{2}\pi n_b \sigma_{ab}^2 \bar{u}_{ab})^{-1}, \tag{95}$$

where σ_{ab} has been introduced previously [Eq. (16)]. The average molecular speed can be defined by $\bar{u}_{ab} = \frac{1}{2}\sqrt{(\bar{u}_a^2 + \bar{u}_b^2)}$.

Equation (95) can be used generally to define an effective collision diameter σ_{ab}. However, the physical significance of this quantity depends on the inter-molecular potential function involved in the collision. The time τ represents, in any case, the time between successive collisions. Often two types of collision are defined. They may be considered to be two limiting cases, namely

(i)   Adiabatic collisions, for which there is a significant change in the phase of the molecular vibration, and

(ii)   Non-adiabatic collisions, in which there is a resulting change in energy corresponding to a quantum transition. They are often referred to as collisions of the first and second kind, respectively.

In optical spectra, the pressure broadening is the result primarily of adiabatic collisions, thus non-adiabatic collisions can be ignored. However, at low frequencies, in the microwave region, for example, adiabatic transitions are much less important and the broadening is the result primarily of transitions from one rotational (or inversion) state to another. Experimental measurements of the width of a microwave absorption then offer a direct method of determining the effective cross-sections for bimolecular interactions. The broadening is usually the result of collision-induced transitions between translation and rotation of the molecule of interest.

Chapter 15

Gas Lasers

As pointed out in Chapter 12, it is the inversion of population that is fundamental for the functioning of lasers (see, for example, Willett in 'Further Reading' at the end of the book). Thus, as indicated in Fig. 33, if an excess (with respect to the Boltzmann distribution at a given temperature) of molecules or atoms can be excited to higher energy levels at a given time, there will be a certain probability of stimulated emission. The resulting electromagnetic emission will then be at a frequency that corresponds to the difference between its initial (higher) energy and that of the final energy involved in the transition. The design of a functioning laser then depends on the choice of an atomic or molecular system with appropriate energy levels and lifetimes, as well as a method to excite, or pump, the emitting species to the desired higher energies.

The method of excitation, or, perhaps better, selection, employed in the ammonia maser was an electric field, as explained in Chapter 12. This example corresponds to a two-level system, as represented in Fig. 37(a). The chosen molecules populate the excited state to yield the necessary population inversion. Their return to the ground state by stimulated emission then produces the radiant energy; in this case, at microwave frequencies. As this principle was extended to

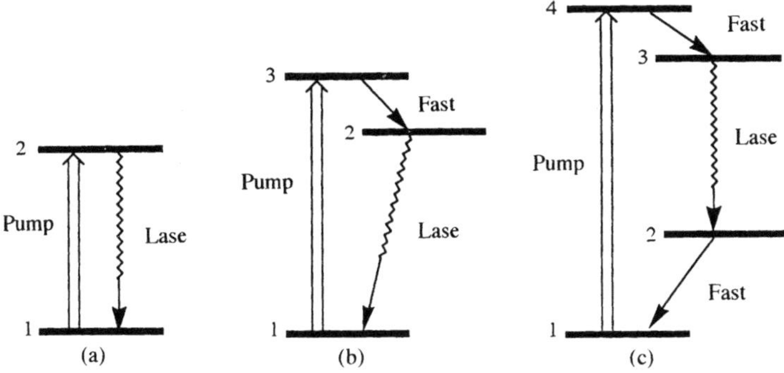

Fig. 37 Typical laser systems: (a) two-level; (b) three-level; (c) four-level

light emission, that is, the laser, more complicated systems were developed. The three-level and four-level laser systems are illustrated schematically in Figs 37(b) and 37(c), respectively. What is most important in these diagrams is the indication of the relative times involved in the various processes. Thus, in Figs 37(b) and 37(c), it is the rapid population of levels 2 and 3, respectively, that is fundamental in the development of laser operation.

Although the first laser was a solid-state device, many, if not most, of the lasers now in use employ gaseous systems [23]. In this case, the pumping operation is usually an electrical discharge in the gas. This excitation can be induced by microwave radiation outside the gas cell or by a high voltage applied between electrodes inside the cell. Pumping is achieved as a result of collisions between the gaseous atoms or molecules and electrons that are produced with relatively high translational energy in the discharge.

In many gas lasers, a mixture of gases is used; namely, that of the lasing gas, N, and a second gas, M. The role of M is only to be excited to M^* by collisions with electrons and to transfer this energy to N by subsequent collisions of the second kind. Thus,

$$N + M^* \rightarrow N^* + M. \tag{96}$$

In the ideal case, M^* has a long-lived, metastable state with an energy that is approximately in resonance with a transition in the lasing species, N.

15.1 THE HELIUM–NEON LASER

One of the first lasers, and still one of the most important, was the helium–neon laser (see Milonni and Eberley in 'Further Reading' at the end of the book). Its operation will now be described briefly, as it can be considered to be a prototype of the gas-phase lasers that are employed in current applications. It is a continuous-wave (CW) laser that is reliable and relatively simple in operation. The mechanism of laser action can be understood by reference to the basic four-level scheme shown in Fig. 37(c). However, a detailed description of the various states and transitions involved is much more subtle and, in fact, not entirely known.

The overall energy diagram for the helium–neon system is shown in Fig. 38. Pumping is accomplished by means of an electrical discharge, usually produced by passing a high voltage between two electrodes. Helium atoms are excited by collisions with the electrons created in the discharge plasma. Of the various excited states resulting from this process, the 2^3S_1 and 2^1S_0 states are metastable. They have relatively long lifetimes because the de-excitation transition to the 1^1S_0 ground state is forbidden by the appropriate selection rules.

The energy-level diagram for neon is considerably more complicated. Here, also, the ground state is $^1S_0$, which arises from the $1s^2 2s^2 2p^6$ electronic configuration. However, the many excited states of interest cannot be correctly described by the usual Russell–Saunders (L–S) coupling scheme. Each electronic

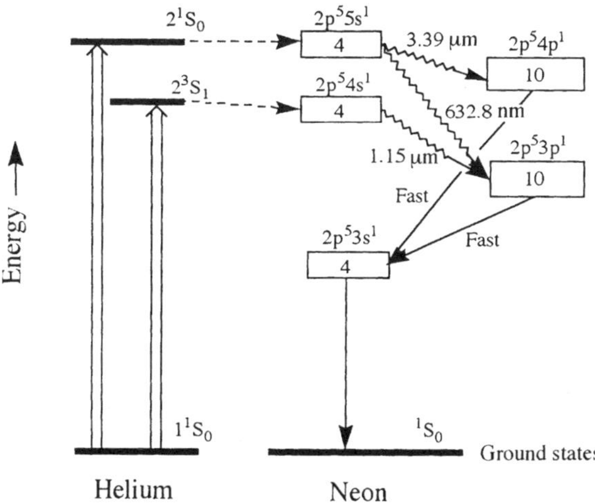

Fig. 38 Transitions in the helium–neon laser

configuration results in a number of states with slightly differing energies. In Fig. 38 each rectangle represents a manifold of four or 10 of such states. The exact description of these states is unimportant for the basic understanding of laser operation.

As indicated in Fig. 36, the 2^3S_1 state of helium has an energy that is close to the energies of the manifold of states resulting from the $2p^54s^1$ configuration of neon. A similar near-resonance is evident between the 2^1S_0 state of helium and the levels of the $2p^55s^1$ configuration of neon. Collisions of the second kind can thus transfer energy relatively easily, as shown in the figure.

The states corresponding to the $2p^54s^1$ and $2p^55s^1$ configurations have lifetimes of the order of 0.1 μs. However, the various states within the manifolds of configurations $2p^54p^1$ and $2p^53p^1$ have shorter lifetimes by approximately one order of magnitude. Thus, their depopulation is relatively rapid. These conditions correspond to those that are ideal to produce the inversion of population necessary for laser operation.

The most useful laser lines produced by the helium–neon system are those around 1.15 μm, in the near-infrared region, and those in the visible region. Of the latter, that at 632.8 nm yields the very strong red-light output that is characteristic of the helium–neon laser. The depopulation of the $2p^53s^1$ levels is assured by collisions with the walls of the discharge tube. Thus, tubes of only a few millimeters in diameter are employed to increase the efficiency of this process. The choice of the laser output is determined by the resonant wavelength of the optical cavity employed in the system, as explained in what follows.

The configuration of a typical helium–neon laser is illustrated in Fig. 39. The composition of the gas mixture is of the order of 10/1 in He/Ne, at a total

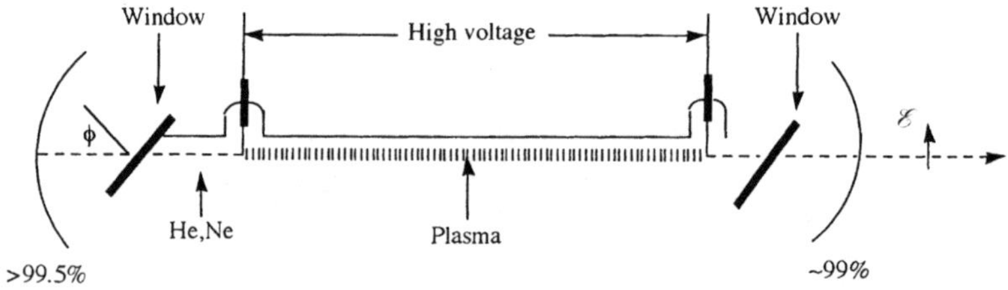

Fig. 39   Configuration of the helium−neon laser

pressure of about 1 Torr. The discharge tube is fitted with glass windows mounted at the Bruster angle. Students often ask why the windows on the laser tube are at such a strange angle with respect to the tube axis. This angle, which is chosen to yield a minimum of light loss, depends on the index of refraction n of the window material and, to a certain extent, on the wavelength of the light. In this application, the Bruster angle corresponds to $\tan \phi = n$, or $\phi \approx 57°$ for glass windows, with light in the visible region.

The optical resonant cavity necessary to build up the signal at the chosen wavelength consists of two spherical mirrors, as shown in Fig. 39. The distance between the two mirrors is carefully adjusted to obtain resonance at the desired wavelength. The reflectivity of the mirror to the right in this figure is slightly reduced in order to obtain the output from the laser system. Within the resonant cavity a number of different modes of oscillation can exist. However, with care, a particular mode, usually the fundamental (TEM_{00}) mode can be chosen, which is plane-polarized. The Bruster-angle configuration of the windows aids this selection considerably. The He−Ne laser constitutes a coherent, essentially monochromatic, polarized light source that now has many very important technological and scientific applications. A number of other gas lasers of this type have been developed and commercialized.

15.2   THE CARBON DIOXIDE LASER

This laser produces high power in the mid-infrared spectral region. Its operation depends on transitions between the various rotation−vibration levels of the CO_2 molecule.

Carbon dioxide is a linear molecule. Thus, with three atoms, it has $3 \times 3 - 5 = 4$ vibrational degrees of freedom, as indicated in Chapter 4. They correspond to three vibrational frequencies: a symmetrical stretching mode, designated $\nu_1(\Sigma_g^+)$, a doubly degenerate bending mode, $\nu_2(\Pi_u)$ and an antisymmetric stretching mode, $\nu_3(\Sigma_u^+)$. Here, again, a knowledge of spectroscopic notation is unnecessary for the understanding of the functioning of the laser. It should be noted, however, that mode ν_1 is coupled, as a result of the anharmonicity of the vibrations, to the first overtone of the bending mode, ν_2.. The result is known as a Fermi resonance. It produces a doublet, as indicated by the bracket on the right in Fig. 40. It will be seen below that this doublet plays an important role in the operation of the CO_2 laser.

As in the He–Ne laser, the initial excitation is provided by an electrical discharge. Nitrogen molecules are excited to the $v = 1$ level, where they remain in a metastable state. Their return to the ground state is forbidden by the vibrational selection rules, as this molecule is non-polar. The energy difference between the excited vibrational state of nitrogen ($v = 1$) and the 00^01 level of carbon dioxide is only 18 cm^{-1}. The exchange of energy through collisions of the second kind is thus relatively easy and the first excited ν_3 level (00^01) becomes preferentially populated. Two lasing transitions are observed at 9.6 and 10.6 µm, as shown. Laser operation is assured by the rapid decay of the Σ_g^+ levels to the first excited state of $\nu_2(\Pi_u)$. A low partial pressure of He is usually added to the gas mixture to aid, via inelastic collisions, the rapid depopulation of this level to the ground state.

The laser output indicated in Fig. 40 is composed of a large number of transitions as a result of the rotational levels of the CO_2 molecule. In this energy diagram, each level for CO_2 represents a vibrational state, upon which is superimposed the series of rotational energies given by $\varepsilon_{\text{rotation}} = (h^2/8\pi^2 I)J(J + 1)$, as explained in Chapter 4. The rotational quantum number $J = 0, 1, 2 \ldots$ and I is the moment of inertia of this linear molecule. The various rotational contributions to a given transition are determined by the selection rules $\Delta J = \pm 1$. In the rotation–vibration spectrum of CO_2 the transitions with $\Delta J = -1$ yield the so-called P-branch lines, while the R-branch lines

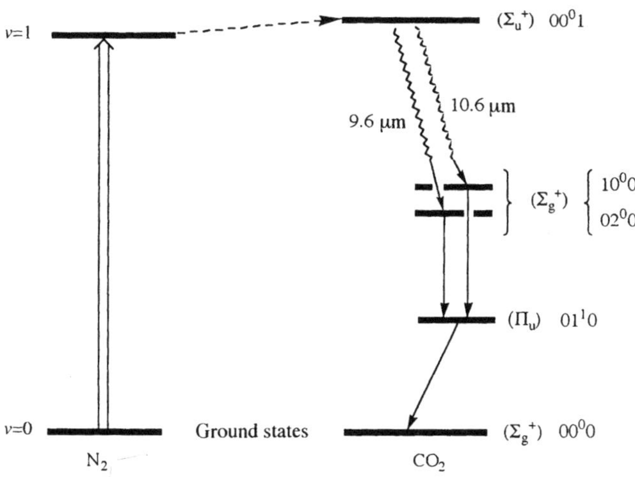

Fig. 40   Transitions in the carbon dioxide laser

are the result of the transitions in which $\Delta J = +1$. The result in the laser application of this molecular system is to produce a large number of closely spaced transitions that can be used to obtain laser output. The particular rotation–vibration laser transition is selected by the optical system, which is resonant at the appropriate wavelength. Often, one mirror, as on the left of Fig. 39, is replaced by a grating to select the desired output. The relative intensities of the different laser signals are determined essentially by the thermal population of the various initial rotational levels.

The CO_2 laser can operate in either the CW or pulsed mode. In the CW mode, powers of 1 kW can be obtained. Because of their high powers, these devices are of extreme industrial importance.

Chapter 16

Gas Plasmas

The term plasma was originally applied by Langmuir in 1928 to an ionized gas in an electrical discharge. This term has now been generalized to include a number of neutral mixtures that contain charged particles. However, the most important of these systems is still the ionized gas. Two particular cases will be considered in this chapter; namely, gaseous plasmas formed under equilibrium conditions and electrical conductivity in non-equilibrium plasmas (see Tanenbaum in 'Further Reading', at the end of the book).

16.1 EQUILIBRIUM PLASMAS

The energy required to remove an electron from a neutral atom or molecule is referred to as its first ionization potential. Some typical values are presented in Table 3, where the energy is expressed in electron volts (eV). It should be noted that with energy equal to kT (where k is the Boltzmann constant), 1 eV corresponds to a temperature of 11 600 K. It is apparent, then, that very high temperatures are necessary to produce significant ionization under equilibrium conditions. An analysis of this question can be made by application of the Boltzmann distribution law [Eq. (I.46)].

In Chapter 4 it was assumed that the electronic partition function was equal to unity. This approximation is acceptable at 'ordinary temperatures' if the ground electronic state is non-degenerate. However, at the temperatures that are attained in plasmas, this value must be modified to take into account the population of the excited electronic states. Their density becomes very great as the ionization energy is approached. Thus, the electronic partition function in this case can be expressed in the form

$$\mathscr{Z}_{\text{electrons}} = \sum_{i=0}^{\infty} g_i \, e^{-\varepsilon_i/kT} \tag{97}$$

$$= g_0 + g_1 \, e^{-\varepsilon_1/kT} + g_2 \, e^{-\varepsilon_2/kT} + \ldots, \tag{98}$$

as the energy of the ground electronic state is taken equal to zero. The series

Table 3   First ionization potentials of some simple molecules

Compound	First ionization potential (eV)
H_2	15.60
N_2	15.51
Cl_2	13.20
Br_2	12.80
O_2	12.50
S_2	10.70
I_2	9.70
CO	14.10
HCl	13.80
HBr	13.20
HI	12.80
NO	9.50
HCN	14.80
SO_2	13.10
N_2O	12.90
H_2O	12.56
NO_2	11.00
H_2S	10.42
NH_3	11.20

represented by Eqs (97) and (98) is not convergent in the mathematical sense. However, in practical applications, use can be made of the fact that the energies of the electronic states all approach that given by the first ionization potential, U_1. Thus, the system can be considered to consist of two energy levels, namely the ground state, with degeneracy g_0, and the essentially ionized state with degeneracy $g_{ion} = \sum_{i=1}^{m} g_i$, where m represents a somewhat arbitrary limit. In any case, $g_{ion} \gg g_0$.

Suppose that N_{mol} represents the number of unionized species and N_{ion} the number ionized. Their ratio is given by

$$\rho \equiv \frac{N_{ion}}{N_{mol}} = \frac{g_{ion}}{g_0} \, e^{-(\varepsilon_{ion}-\varepsilon_0)/kT}, \tag{99}$$

where g_0 and g_{ion} are the statistical weights; $\varepsilon_0 = 0$ and $\varepsilon_{ion} \approx U_1$ are the corresponding energies. For simplicity, it is assumed that the system is described by only these two energy levels. Then the percentage of ionized species is expressed as

$$\alpha = \frac{N_{ion}}{N_{ion} + N_{mol}} = \frac{\rho}{1 + \rho}. \tag{100}$$

Thus, from Eq. (99), $\rho = 1$, or $\alpha = 50\%$, at the temperature $T_{1/2} =$

$U_1/[k \ln(g_{\text{ion}}/g_0)]$, a temperature well below that which corresponds to the ionization potential.

The ratio g_{ion}/g_0 appearing in Eq. (99) can be estimated if it is realized that it is the relatively free electrons produced by the ionization process that characterize the plasma. The electrons, being much lighter than either the positive ions or the neutral atoms, move much more rapidly and the transfer of their kinetic energy to the heavier species is very inefficient. It is effectively the translational partition function for the electrons [see Eq. (I.78)] that determines the quantity

$$\frac{g_{\text{ion}}}{g_0} = \frac{1}{n_{\text{ion}}} \left(\frac{2\pi m_e kT}{h^2}\right)^{3/2}, \tag{101}$$

where n_{ion} is the number of ions per unit volume. Then, the ratio of ionized species to unionized species is given by

$$\rho = \frac{1}{n_{\text{ion}}} \left(\frac{2\pi m_e kT}{h^2}\right)^{3/2} e^{-U_1/kT}, \tag{102}$$

which is one form of the Saha equation. If the 'electron gas' is considered to be ideal, $n_{\text{ion}} = P/kT$, and the equilibrium constant for the ionization process, expressed in terms of pressures, becomes

$$K_P = \frac{\alpha^2 P}{1 - \alpha} = kT \left(\frac{2\pi m_e kT}{h^2}\right)^{3/2} e^{-U_1/kT}. \tag{103}$$

This result shows that extremely high temperatures are necessary to produce a significant ionization of a gas under equilibrium conditions. However, the ionization is enhanced by very low pressures, as in a high-vacuum system.

If the pressure P is expressed in centimeters of Hg (Torr) and the first ionization potential is given in electron volts (see Table 3), then the degree of ionization can be calculated from the practical relation

$$\frac{\alpha^2 P}{1 - \alpha} = 2.5 \times 10^{-4} T^{5/2} e^{-1.16 \times 10^4 U_1/T}. \tag{104}$$

In deriving Eqs (103) and (104) it was assumed that the electrons do not have a significant effect on the pressure.

It is of interest to apply Eq. (104) to a particular example, say, hydrogen atoms. The results are shown in Fig. 41 for various pressures. It is perhaps surprising that the transition from the unionized state to the totally ionized state occurs within a very limited temperature range. This result is suggestive of a phase transition; and it is for this reason that Crookes in 1879 referred to the plasma as the fourth state of matter. In fact, it is a rather strange state because of the Coulombic interactions between the charged particles, which are very long-ranged (see Chapter 6). The notion of a 'collision' then becomes difficult to define. Hence, in a plasma an electron interacts simultaneously with many of its

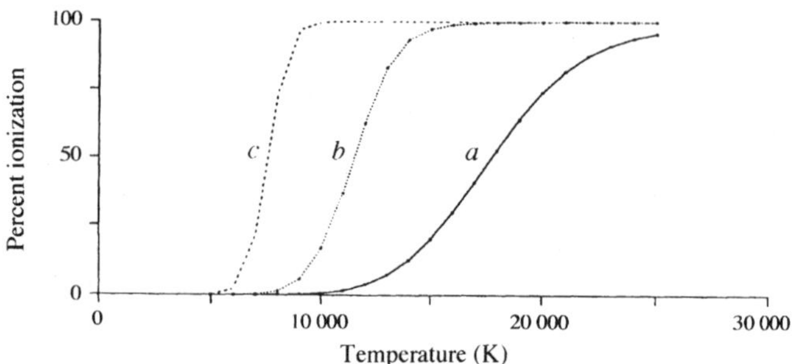

Fig. 41　Fraction of ionized species as a function of temperature, as calculated from the Saha equation [Eq. (104)]: a: $P = 760$ Torr (1 atm); b: $P = 1$ Torr; c: $P = 10^{-3}$ Torr

neighbors, resulting in a certain cohesive property analogous to that of a jelly—hence, the term plasma.

16.2　ELECTRICAL CONDUCTION

It is apparent from the above discussion that under equilibrium conditions very high temperatures are necessary to produce a significant concentration of electrons. Such temperatures are characteristic of stars, although they are not easy to achieve under laboratory conditions here on earth. The shock tube technique described in Chapter 4 is, however, a relatively convenient method of producing very high temperatures, although the time available for observations is short. The more usual way of producing highly ionized gases is an electrical discharge. The passage of an electric current results in a non-equilibrium situation in which the electrons are usually at a very high temperature with respect to the heavier species. Visible examples of electrical discharges in gases include flashes of lightning, or artificial sources such as light emitted by commercial neon tubes, arc welding machines, etc.

The basic approach to a description of a plasma assumes that it is, in fact, the electrons that play the dominant role. It is useful to define the electron collision frequency, ν, which is the number of collisions per second that the average electron undergoes with the heavier particles in the plasma. The basic equation of motion, which is attributed to Langevin, can be simplified in the form

$$\mathbf{J} = \sigma_0(\mathscr{E} + \mathbf{u} \times \mathbf{B}_0), \tag{105}$$

where the dc conductivity of the plasma is given by $\sigma_0 = n_{\text{ion}}e^2/m_e\nu$ and the current density is equal to $\mathbf{J} = -n_{\text{ion}}e\mathbf{u}$. Here, $\mathbf{u}$ is the average electron velocity and $\mathscr{E}$ and $\mathbf{B}_0$ are respectively the electric and magnetic fields acting on the electron. Equation (105) is often referred to as the generalized Ohm's law.

It is worth noting that in the absence of a magnetic field, scalar multiplication of Eq. (104) by **J** yields

$$\mathbf{J}^2 = \sigma_0 \mathbf{J} \cdot \mathscr{E}. \tag{106}$$

Clearly, $\mathbf{J} \cdot \mathscr{E}$ is the rate at which work per unit volume is done on the electric current, while

$$\frac{1}{\sigma_0} \mathbf{J}^2 = \rho_0 \mathbf{J}^2, \tag{107}$$

where ρ_0 is the resistivity, expresses the heating of the gas by the current. To maintain a steady state the heat generated must be radiated as fast as it is produced.

Returning now to the generalized Ohm's law, with $\mathbf{u} = -\mathbf{J}/(e n_{\mathrm{ion}})$, Eq. (105) leads to

$$\mathbf{J} = \sigma_0 \left(\mathscr{E} - \frac{1}{e n_{\mathrm{ion}}} \mathbf{J} \times \mathbf{B}_0 \right). \tag{108}$$

It is convenient to choose the coordinates so that $\mathbf{B}_0$ is along the z axis. Then the development of Eq. (108) yields the relations

$$J_x = \sigma_0 \left(\mathscr{E}_x - \frac{B_z}{e n_{\mathrm{ion}}} J_y \right), \tag{109}$$

$$J_y = \sigma_0 \left(\mathscr{E}_y - \frac{B_z}{e n_{\mathrm{ion}}} J_x \right) \tag{110}$$

and

$$J_z = \sigma_0 \mathscr{E}_z. \tag{111}$$

By combining Eqs (109)–(111), a vectorial expression for the current density can be written in the form

$$\mathbf{J} = \boldsymbol{\sigma} \cdot \mathscr{E}, \tag{112}$$

where $\boldsymbol{\sigma}$ is a 3×3 matrix (an antisymmetric tensor). With the application of sum algebra it is found that this matrix, which is known as the conductivity tensor, takes the form

$$\boldsymbol{\sigma} = \sigma_0 \begin{pmatrix} \varkappa & -\varkappa B_z / e n_{\mathrm{ion}} & 0 \\ \varkappa B_z / e n_{\mathrm{ion}} & \varkappa & 0 \\ 0 & 0 & 1 \end{pmatrix} = \begin{pmatrix} \sigma_\perp & \sigma_{\mathrm{H}} & 0 \\ -\sigma_{\mathrm{H}} & \sigma_\perp & 0 \\ 0 & 0 & \sigma_0 \end{pmatrix}, \tag{113}$$

where

$$\varkappa \equiv \left[1 + \frac{B_z^2}{e^2 n_{\text{ion}}^2} \right]^{-1}.$$

In Eq. (113), which was derived above, the quantity $\sigma_\perp = \varkappa \sigma_0$ is called the perpendicular conductivity, as it determines current flow in the direction of the electric field $\mathscr{E}$, when it is perpendicular to the direction of the magnetic field B_z. The off-diagonal components of the conductivity tensor, $\sigma_{\text{H}} = \pm \varkappa \sigma_0 B_z / e n_{\text{ion}}$, determine what is known as Hall conductivity. They govern the current flow in the direction perpendicular both to $\mathscr{E}$ and to $\mathbf{B}$—the latter being along the z axis in this case. It is apparent that $\varkappa$ decreases very rapidly with an increasing magnetic field. Thus, the first two diagonal elements of the conductivity tensor, $\sigma_\perp$, decrease rapidly with an increasing magnetic field and very little current can be driven across the lines of an applied magnetic field. The Hall conductivity also decreases as the strength of the magnetic field increases, although the effect is less rapid. Finally, it is evident that the quantity σ_0 determines the flow of current in the z direction, that of the magnetic field.

References

[1] H. Eyring, *J. Chem. Phys.* **3**, 107 (1935)

[2] H. R. Mayne and J. P. Toennies, *J. Chem. Phys.* **70**, 5314 (1979); *J. Chem. Phys.* **75**, 1794 (1981)

[3] George Porter and Franklin J. Wright, *Z. Electrochem.* **56**, 782 (1952)

[4] J. G. Anderson, W. H. Brune and M. H. Proffitt, *J. Geophys. Res. D: Atmos.* **94**, 11465 (1989)

[5] J. G. Anderson, W. H. Brune, S. A. Lloyd, D. W. Toohey, S. P. Sander, W. L. Starr, M. L. Loewenstein and J. R. Podolske, *J. Geophys. Res. D: Atmos.* **94**, 11480 (1989)

[6] George Porter and Michael R. Topp, *Nature* **220**, 1228 (1968)

[7] M. Dantus, M. J. Rosker and A. H. Zewail, *J. Chem. Phys.* **87**, 2395 (1987)

[8] M. Rosker, M. Dantus and A. H. Zewail, *Science* **241**, 1200 (1988)

[9] A. Farkas, *Orthohydrogen, Parahydrogen and Heavy Hydrogen*, Cambridge University Press, 1935

[10] R. C. Miller and P. Kusch, *Phys. Rev.* **99**, 1314 (1955)

[11] O. Stern and W. Gerlach, *Ann. Phys.* **74**, 673 (1924)

[12] J. P. Gordon, H. J. Zeiger and C. H. Townes, *Phys. Rev.* **99**, 1264 (1955)

[13] R. R. Herm, R. Gordon and D. R. Herschbach, *J. Chem. Phys.* **41**, 2218 (1964)

[14] D. H. Herschbach, *Adv. Chem Phys.* **10**, 319 (1968)

[15] A. E. Grosser and R. B. Bernstein, *J. Chem. Phys.* **43**, 1140 (1965)

[16] R. Götting, H. R. Mayne and J. P. Toennies, *J. Chem. Phys.* **85**, 6396 (1986)

[17] S. H. Bauer, *Science* **141**, 867 (1963)

[18] E. F. Greene and D. F. Hornig, *J. Chem. Phys.* **21**, 617 (1953)

[19] A. G. Bell, *Am. J. Sci.* **20**, 305 (1980); *Phil. Mag.* **11**, 510 (1881)

[20] P. V. Slobodskaya, *Izv. akad. nauk. SSSR.* **12**, 656 (1948)

[21] M. E. Jacox and S. H. Bauer, *J. Phys. Chem.* **61**, 833 (1957)

[22] T. H. Cottrell, I. M., McFarlane and Read, A. W., *Trans. Faraday Soc.* **63**, 2093 (1967)

[23] J. D. Rigden and A. D. White, *Proc. IRE* **50**, 7 (1962)

Further Reading

Atkins, P. W., *Physical Chemistry*, Third Edition, Oxford University Press, Oxford, UK, 1978. A very popular physical chemistry textbook.

Bradley, John N., *Shock Waves in Chemistry and Physics*, Methuen & Co. Ltd, London, 1962.

Cambel, Ali Bulent, Duclos, Donald P. and Anderson, Thomas P., *Real Gases*, Academic Press, New York, 1963. This monograph presents an engineering approach to the subject.

Garrett, C. G. B., *Gas Lasers*, McGraw-Hill Book Company, New York, 1967. This book, like the one by Willett, presents in some detail the development of this important topic.

Golden, Sidney, *Elements of the Theory of Gases*, Addison-Wesley Publishing Company, Inc. Reading, Massachusetts, 1964. This work presents in simple terms the theoretical aspects of the subject.

Green, E. F. and Toennies, J. P., *Chemical Reactions in Shock Waves*, Edward Arnold Ltd, London, 1964. An English translation of an earlier edition in German.

Hildebrand, Joel H., *An Introduction to Molecular Kinetic Theory*, Reinhold Book Corporation, New York, 1963. This little book covers some of the more elementary aspects of the kinetic theory of gases.

Hirschfelder, J. G., Gurtiss, C. F. and Bird, R. B., *Molecular Theory of Gases and Liquids*, John Wiley & Sons, Inc., New York, 1954. This work is certainly the most complete treatment of the dynamics of molecules in the gaseous and liquid states. It is indispensable for advanced graduate students and researchers in the field. It is not, however, easy reading.

Hollas, J. Michael, *Modern Spectroscopy*, John Wiley & Sons, Chichester, 1987. This book provides a very good introduction to atomic and molecular spectroscopy.

Kennard, E. H., *Kinetic Theory of Gases*, McGraw-Hill Book Company, Inc., New York, 1938. This book is one of the classic advanced textbooks devoted to the kinetic theory of gases.

Laidler, K. J., *Chemical Kinetics*, McGraw-Hill Book Company, New York, 1950. The classical textbook on the kinetics of chemical reactions.

Levine, Ira N., *Molecular Spectroscopy*, John Wiley & Sons, New York, 1975. This book on molecular spectroscopy is more advanced than the one by Hollas.

Levine, Raphael D. and Berstein, Richard B., *Molecular Reaction Dynamics and Chemical Reactivity*, Oxford University Press, New York, 1987. This book covers some of the more recent developments—in particular, the analyses of crossed-beam experiments and molecular energy transfer. It is not very readable by undergraduate students.

Maitland, G. C., Rigby, M., Smith, E. B. and Wakeham, W. A., *Intermolecular Forces*, Oxford University Press, Oxford, 1981. A more recent treatment of the specific subject of intermolecular forces than that found in the book by Hirschfelder *et al*.

McQuarrie, Donald M., *Statistical Mechanics*, Harper Collins Publishers, New York, 1976. An advanced treatment of statistical mechanics, with many interesting applications.

Milonni, Peter W. and Eberly, Joseph H., *Lasers*, John Wiley & Sons, New York, 1988. This is a more general, and more recent, work in this field than the books by Garrett and by Willett.

Minkoff, G. J. and Tipper, C. F. H., *Chemistry of Combustion Reactions*, Butterworths, London, 1962.

Moore, Walter J., *Physical Chemistry*, Prentice-Hall, Inc., Englewood Cliffs, NJ, 1955. This book was the first of two most popular physical chemistry textbooks. The sections on gases is a useful, and very readable, introduction to the subject.

Partington, J. R., *An Advanced Treatise on Physical Chemistry*, Volume One, Longmans, Green and Co., London, 1949. This first volume in the series is devoted to the properties of gases. It treats the basic theory and presents many of the classical applications.

Pauling, Linus and Wilson, E. Bright, Jr., *An Introduction to Quantum Mechanics*, The McGraw-Hill Book Company, New York, 1935. This is the classic work on chemical applications of quantum mechanics. The many books that have followed are firmly based on, if not copied from, this one.

Present, R. D., *Kinetic Theory of Gases*, McGraw-Hill Book Company, Inc., New York, 1958. An advanced textbook on the kinetic theory of gases.

Rushbrook, G. S., *An Introduction to Statistical Mechanics*, Oxford University Press, Oxford, UK, 1949. One of the classical introductory books on this subject.

Spitzer, Lyman, Jr., *Physics of Fully Ionized Gases*, Interscience Publishers, Inc. New York, 1956. Similar to the book by von Engel.

Tanenbaum, B. Samuel, *Plasma Physics*, McGraw-Hill Book Company, New York, 1967. See the earlier works by von Engel and by Spitzer.

von Engel, A., *Ionized Gases*, Oxford University Press, Oxford, 1955. This work presents a development of the subject of gas plasmas, which is only briefly treated in the present volume.

Willett, C. S., *An Introduction to Gas Lasers: Population Inversion Mechanisms*, Pergamon Press, Oxford, 1974.

Zewail, Ahmed (ed.), *The Chemical Bond. Structure and Dynamics*, Academic Press Limited, London, 1992. This relatively recent book, written in honor of Linus Pauling, offers a fascinating description of the evolution of the present picture of the chemical bond. The second part of the work is devoted to the dynamics of the chemical bond as studied both experimentally and theoretically. It should be required reading for all students and researchers in physics and chemistry—as well as those in that no-man's land of chemical physics.

Index